HOW TO WRITE AND ILLUSTRATE A SCIENTIFIC PAPER SECOND EDITION

This Second Edition of *How to Write and Illustrate a Scientific Paper* will help both first-time writers and more experienced authors, in all biological and medical disciplines, to present their results effectively. Whilst retaining the easy-to-read and well-structured approach of the previous edition, it has been broadened to include comprehensive advice on writing compilation theses for doctoral degrees, and a detailed description of preparing case reports. Illustrations, particularly graphs, are discussed in detail, with poor examples redrawn for comparison. The reader is offered advice on how to present the paper, where and how to submit the manuscript, and finally, how to correct the proofs. Examples of both good and bad writing, selected from actual journal articles, illustrate the author's advice – which has been developed through his extensive teaching experience – in this accessible and informative guide.

BJÖRN GUSTAVII has been teaching courses in scientific writing for doctoral (Ph.D.) students in medicine for 25 years. He brings his personal experience to this book, both from writing more than 100 of his own research papers and from his work as a journal editor.

How to
Write and
Illustrate

Scientific Papers

Björn Gustavii

Second Edition

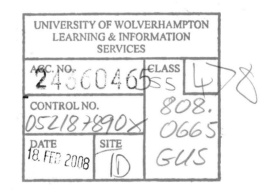

CAMBRIDGE
UNIVERSITY PRESS

CAMBRIDGE UNIVERSITY PRESS

Cambridge, New York, Melbourne, Madrid, Cape Town, Singapore, São Paulo

Cambridge University Press
The Edinburgh Building, Cambridge CB2 8RU, UK

Published in the United States of America by Cambridge University Press, New York

www.cambridge.org
Information on this title: www.cambridge.org/9780521703932

Originally Published in English by Studentlitteratur Lund, Sweden 2000
First published by Cambridge University Press 2003
Reprint 2005,2006
Second Edition Published 2008

Printed in the United Kingdom at the University Press, Cambridge

A catalogue record for this publication is available from the British Library

ISBN 978-0-521-87890-6 hardback
ISBN 978-0-521-70393-2 paperback

Contents

Preface

Dear Novice Writer,

When I was in your shoes and preparing my first paper, I consulted a book on how to write. I found there a sentence encouraging the reader to stand in boiling water for an hour before doing the analysis:

> After standing in boiling water for an hour, examine the contents of the flask.

I had a pretty good idea of what was wrong with the sentence but, at the time, I couldn't figure out how to revise it, and the author didn't tell me. Now I can. If, an hour later, you are still alive:

> Place the flask in boiling water for an hour, then examine its contents.

So, in this book, every unfortunate example is followed by an improved version. Good examples are provided with appropriate bibliographic references. Bad ones, however, are presented with references expunged.

Some examples were taken from manuscripts in preparation, presented by participants in my courses on scientific writing. I have been holding such courses for doctoral (Ph.D.) students in medicine since 1980. Other specimens are from manuscripts submitted for publication. They were collected when I served as an editor of *Acta Obstetricia et Gynecologica Scandinavica* from 1986 to 1994. Yet others are from published material.

From class discussions I have learned what candidates want to know. Based on this information, some chapters are more detailed than others, such as the one on how to prepare graphs.

The current edition contains a new and comprehensive chapter on doctoral (Ph.D.) theses. Numerous other changes also appear in this edition, for example, instructions for making drawings and a description of preparing case reports.

Finally, don't accept all my suggestions, because there is no ultimate truth regarding how to write a paper – as I mistakenly believed when I was a bit younger.

Good luck, my friend.

Björn Gustavii

Acknowledgments

I thank the following persons, who have read all or parts of the manuscript of the second edition, for their advice and criticism.

Per Bergsjø, Norway
Carol Brimley-Norris, Finland
Joy Burrough-Boenisch, UK
Johan Ljungqvist, Sweden
Helen Sheppard, Sweden
Ray Williams, UK
Pål Wölner-Hanssen, Sweden

Special thanks to Tomas Söderblom, a layperson, who read the manuscript for intelligibility; Richard Fisher, who corrected the language; and Eva Dagnegård, who redrew the graphs and prepared the electronic manuscript.

1
Basic rules of writing

Winston Churchill was sitting at his desk, working on his epic about World War II, when his private secretary entered the room. Churchill had reached the Blitz – the German air strikes against London. His staff of researchers had earlier produced a 150-page brief on the raids. The secretary had been asked to cut it down to about two and a half pages and, after having "worked like stink," he could now proudly hand over the condensed version.

Churchill took out his red pen and started to edit. "All my sloppy sentences were tightened up and all my useless adjectives obliterated," the secretary tells us in a documentary made about 50 years later (Bennet 1992). In the midst of it all, Churchill said gently, "I hope you don't mind me doing this?" The secretary answered, "Thank you, Sir – you are giving me a free lesson in writing plain English."

Brevity

We should emulate Churchill by excluding every nonessential word. Professional writers do it that way. Brevity is an elementary rule of all writing, not only to save valuable publication space, but also because verbose writing obscures meaning and wastes the reader's time and patience. And that is also the essence of the next basic rule.

Logic and clarity

To convey information is above all a matter of logic and clarity. What you want to say should be so arranged that the reader can follow your argumentation step by step. Moreover, your sentences should be so clear and easily understood "that the reader forgets that he is reading and knows only that he is absorbing ideas" (Baker 1955).

Now to the importance of making the manuscript physically attractive. Here is an illustrative example.

Clean typing

Paul Fogelberg, editor of a Finnish scientific journal, was one of the teachers at a course on scientific writing. Late one evening, he told us, he was perusing a manuscript in which only half of the letter "a" was legible. Page after page, that half-letter pursued him until eventually he began to feel vaguely that this must be something directed at him personally.

I didn't see Fogelberg again until 12 years later at a meeting of editors. I mentioned the damaged typeface, without really expecting that he would remember it. But he replied instantly, "It wasn't damaged. Much worse – it wasn't cleaned."

Does a dirty typeface of a mechanical typewriter, or an error related to electronic word processing, really matter? Yes, because editors know from experience that there is a close relationship between a poorly prepared manuscript and poor science. So make sure your manuscript looks carefully prepared; it may influence editors and referees in your favor.

2

Comments on scientific language

A MEDLINE search showed that no fewer than 90 percent of papers listed in *Index Medicus* in 1999 were written in English, compared with 53 percent in 1966 (the year MEDLINE started). The saying "Publish in English or perish" must therefore be taken seriously. Regrettably, this means that many authors are obliged to write in a language other than their native tongue – with all that this can entail. Here I will share with you an episode from my own experience as a non-native writer of English.

English as a foreign language

My first paper published in English was initially written in Swedish and then translated into English by a professional translator. "Brilliant," I thought when I saw the translated version. But when my supervisor read it, he shook his head and said, "Try to write directly in English!" "Gosh," I said to myself, thinking of my poor grades in English at school, "I'll never, ever be able to do that."

But I decided to try and consulted the textbooks, which advised me to read writers of fine English, such as Gibbon and his *Decline and Fall of the Roman Empire*. I bought the book (running to

3616 pages in three volumes!) but could find neither the time nor the interest to read it.

Instead, I subscribed to the American weekly magazines *Newsweek* and *Time*. As they often cover the same topics, the reader is given the opportunity to learn twice, in different words, about the same issues. I have found this very instructive.

I have also found another method that has served me well. When I have to tackle a new topic, I read leading English-language publications, underline useful phrases and words, and then create a list of the terms for each section (Introduction, Methods, etc.). I noticed, however, that I seldom had to consult my list. During the process of making the list, the brain seemed to have retained what I had read and written.

I have hardly ever submitted a manuscript in English without asking a linguist to look at it. Ideally, those correcting English ought to be persons who: (1) not only are native speakers of English but also live in your country and speak its language; (2) return to their native country at least once a year to refresh their English; and (3) have a knowledge of scientific writing. Correctors fulfilling these criteria are a rare species. Many authors therefore have to rely on English-speaking persons who, for instance, happen to be working in their department or laboratory. That may not be so bad, after all, because these persons are no doubt acquainted with your field of research. But you must be aware that native-English-speaking researchers do not necessarily write good English – just as not all Swedish researchers are good at Swedish.

I return to my early paper, translated from Swedish into English. On rereading it 30 years later, I found to my embarrassment that it didn't express exactly what I meant to say, though I found the style elegant. However, even clumsy writing would have been better than this, had it conveyed the information accurately.

Why are papers in biomedicine often almost unintelligible? Maybe an editorial in *The Lancet* (1995) had the answer when it claimed that authors of scientific papers often write more to please the editor than to inform the reader. They dare not depart from the traditional style for fear of having their work rejected.

Another mistake commonly committed by beginners is the compulsion to be "complete." Charlie Chaplin had something to say about that.

Follow the "leitmotif"

The video film *Unknown Chaplin* (Brownlow and Gill 1983) shows unused sequences from Chaplin's productions. Some of them are far funnier than those actually included in the final versions of his films. Why were they excluded? Chaplin gives the answer in his autobiography (Chaplin 1973). "If a gag interfered with the logic of events, no matter how funny it was, I would not use it." You are thus recommended to do as Chaplin did and resist the urge to include every item of evidence obtained. In other words, do not include observations that depart from the main theme – no matter how interesting these may seem to be (you will probably find space for them elsewhere, or they could give rise to hypotheses to be tested in future studies). However, if such information cries out to be mentioned, you can insert it parenthetically – as I did in the previous sentence.

Researchers are often short of time. I once heard of a scientist who only had time to read papers while driving to work! That is one reason for keeping a paper short; another is that superfluous words obscure the meaning.

Verbosity

In the following paragraph, adapted from Kesling (1958), 36 of the 53 words can be omitted:

> Our research, designed to test the fatal effects of PGF2α on dogs, was carried out by intravenously introducing the drug. In the

experiments, a relatively small quantity, 30 mg, was administered to each animal. In each case, PGF2α proved fatal; all 10 dogs expiring before a lapse of five minutes after the injection.

Seventeen words are enough:

> Intravenous injection of 30 mg prostaglandin PGF2α to each of ten dogs killed them within five minutes.

"Omit needless words!" is Rule 17 in Strunk and White's *The Elements of Style* (2000). In the introduction to the third edition of the book, E. B. White, a pupil of Strunk, tells us that his teacher omitted so many needless words in his course in English that he would have been left with nothing more to say at the end of his lesson if he had not used a simple trick: he uttered every sentence three times, "Omit needless words! Omit needless words! Omit needless words!"

But do not go too far. The telegraphic style of the following sentence taken from *Contraception* must be a riddle to a non-specialist:

> Young mature Sprague Dawley rats (200 g) (Charles River Italia) were [used].

What do "young" and "mature" mean? What do "Sprague Dawley" and "Charles River Italia" stand for? And did all the rats weigh exactly 200 g? The average reader is probably better served by this:

> The rats used in this experiment were obtained from Charles River Breeding Laboratories and were derived from the Sprague Dawley strain. The animals were sexually mature, 100 days old, and weighed 190 to 215 g.

He/she

Most writers no longer use male pronouns (*he, his, him*) to denote both males and females. Does this mean that our language is less sexist now? No. Instead, we have got constructions such as *he/she* or *s/he*, which hardly solve the problem, but rather emphasize it. Here is an example from a manuscript:

> Each patient was interviewed at the out-patient unit that s/he belonged to.

How to avoid constructions like this? The simplest way is often to use the plural:

> All *patients* were interviewed at the out-patient unit *they* belonged to.

On the odd occasion where the use of the plural seems impossible, reword the sentence or try to remove the pronoun. For example, in the following, the pronoun *their* could be removed.

> I submitted the manuscript to the editor for their consideration.

Only when all else fails, use the less awkward form *he or she*. Finally, I must relate an anecdote by Sheila McNab (1993).

> In a serious road accident a father was killed and his son seriously injured. When the boy was later brought into the hospital operating theatre, the surgeon blanched and exclaimed, "I can't operate on this boy, he is my son!"

If you were unable to realize immediately that the surgeon was the boy's mother, you may have something to think about. When I tested this anecdote on my graduate students, one male student could find only one answer: the man who had died was the stepfather!

Active or passive voice

Previously, scientists were obliged by tradition to use the passive voice. The use of first-person pronouns (*I* or *we*) was seen as pretentious, even impolite. Not so now. Scientists of today dare do what Watson and Crick, back in 1953, had the courage to do in the opening phrase of their classic on the structure of DNA – and say *we*:

> We wish to suggest . . . ,

which is more direct, easier to read, and shorter than the passive:

> In this letter a suggestion is made . . .

Below is another example, drawn from *New Scientist* (1993). Its former editor, Bernard Dixon, found the following sentence in a manuscript submitted:

> The mode of action of anti-lymphocytic serum has not yet been determined by research workers in this country or abroad.

Dixon replaced it with:

> We don't know how anti-lymphocytic serum works.

"He was quick to telephone me," Dixon recalls, "complaining about editorial interference. [. . .] How could a magazine as prestigious as *New Scientist* change an author's meaning in such a cavalier fashion? But, I replied, we had not altered his meaning. We had simply made a sentence more readable and direct – and cut it to a third of its original length."

However, in methods and results sections the passive voice is generally more effective. It emphasizes the action rather than the person performing the action. Thus, the active form:

> I stopped cell growth with colchicine

has no real advantage over the passive:

> Cell growth was stopped with colchicine

since nobody cares who performed the act. And further, when there are several authors, the *we* in:

> We stopped cell growth with colchicine

is probably not true – unless the authors each added a portion! Thus active and passive voices both have their place in scientific writing.

Tense

Only two tenses are normally used in scientific writing: present and past (Day 1995; Day and Gastel 2006). *Present tense* is used for established knowledge (including your own published findings), *past tense* is used for the results that you are currently reporting.

Most of the abstract section describes your own present work; it is referred to in the past tense. Much of the introduction section emphasizes previously established knowledge; given in the present tense. Here is an example (Dembiec *et al.* 2004; emphases are mine):

> INTRODUCTION
> Tigers **are** often transported [but] the effect of transfer on them has not yet been documented [2]. . . .

The methods and results sections describe what you did and found; they appear in the past tense:

> METHODS
> We **simulated** transport by relocating five tigers in a small individual transfer cage. . . .
>
> RESULTS
> Average respiration rate of all tigers **increased**. . . .

Finally, in the discussion section, where you compare established knowledge with your own findings, you normally see-saw back and forth between present and past tense – even in the same sentence.

Noun clusters and modifiers

In *USA TODAY* (October 13, 1992), I saw this:

> Pig liver transplant woman dies

As a newspaper headline this phrase is acceptable. It is intelligible and unambiguous; and the cramped space makes it necessary. But in a scientific paper, such a sentence would have looked ridiculous. Here, it has been written out in full:

> The woman with a transplanted pig liver has died

The following phrase, quoted from *Contraception*, may be entirely and immediately intelligible to an expert in the field:

> Rabbit anti-mouse spleen cell serum . . .

But researchers not working in that field might wonder to which animal the spleen had belonged. The writer could have saved some readers a little trouble if he had written:

> Anti-mouse serum of rabbits immunized with cells of mouse spleen . . .

However, it is quite acceptable to couple a few nouns and modifiers as long as it is crystal-clear what you mean and as long as the reader can grasp the string of words at first reading, as in this example from a methods section (Mehrotra *et al.* 1973):

> Colony bred female albino rats . . .

and this used as a subheading (Gardiner *et al.* 1980):

> Anaesthetized spontaneously breathing guinea pig

Prevalence and incidence

The words *prevalence* and *incidence* are said to be among the most misused terms in biomedical reports. *Prevalence* refers to the total number of cases of a disease or condition existing at a specific time. *Incidence* refers to the number of new cases that develop over a specific time. In the following example from *Newsweek* (Begley 1996), the prevalence is 200 000 and the incidence 12 000.

> Each year as many as 12 000 Americans join the more than 200 000 who already live with paralyzing spinal-cord injuries.

Avoid the use of "respectively"

Respectively obliges the reader to stop and reread the sentence, as in the following, seen in a manuscript under preparation:

> Phytate reduction in wheat, rye, barley with and without hulls incubated with 40 g water/100 g cereal for 24 hours at 55 °C was 45, 56, 48 and 77%, respectively.

The version below is direct and permits the reader to proceed (revised text in boldfaced italics):

> After incubation with 40 g water/100 g cereal for 24 hours at 55 °C, ***phytate reduction in wheat was 45%; in rye, 56%; in barley with hulls, 48%; and in barley without hulls, 77%.***

The "and/or" construction

The expression *and/or* disrupts the textual flow, as in this example:

> The effect of intravenous streptokinase and/or oral aspirin . . .

which the reader would have found easier if it had read:

> The effect of intravenous streptokinase, oral aspirin, or both . . .

A closer look at the text often reveals that *and/or* can be replaced by *and* (*The ACS Style Guide* 1977):

> Our goal was to confirm the presence of the alkaloid in the leaves and/or roots,

or by *or* (de Looze 2002):

> Confidential information can only be given to the patients and/or close relatives.

The construction *and/or* has no place in scientific writing.

Unnecessary hedging

Hedging is a way of saying "maybe" more than once. Two or more hedges can drain all force from a sentence. The eminent writer in the cartoon replaced seven hedges ("seems," "not inconceivable," "suggest," "may," "indicate," "possible," "probably") with just one: "think." One hedge is always enough.

Figure 2.1 The author seen in the figure thought twice before presenting his message. (Redrawn, with permission, from Majewski 1994.)

How to Write and Illustrate a Scientific Paper

How old is young?

Consider this title of a paper:

> Herniography in younger women with unclear groin pain

The abstract of this paper stated that the women were under 40 years of age. Well, viewed from my age these women were young, but the average readers would probably have appreciated a more precise definition of their age, as in this example (Sundby and Schei 1996):

> Infertility and subfertility in Norwegian women aged 40–42. Prevalence and risk factors.

In other cases, an age range can be defined by a specific term, as in this title (Gold *et al.* 1996):

> Effects of cigarette smoking on lung function in adolescent boys and girls

You are advised to use specific terminology, when available, to report subjects' ages. Here are the age groups recommended by MEDLINE, as of January 2007:

All infants	birth–23 months
All children	0–18 years
All adults	19+ years
Newborn	birth–1 month
Infant	2–23 months
Preschool child	2–5 years
Child	6–12 years
Adolescent	13–18 years
Adult	19–44 years
Middle aged	45–64 years
Middle aged + aged	45+ years
Aged	65+ years
80 and over	80+ years

Don't ask me why the "adult" category excludes persons aged 18 years. Note that *young* is not listed – it is undefinable.

Persons below adult age may be referred to as *boys* and *girls*. For adults, *men* and *women* are the correct terms.

Avoid synonyms to achieve elegant variation

In the list of abbreviations of a manuscript under preparation I found this (boldface italics mine):

> C_{max}: maximum plasma **concentration** achieved.
> T_{max}: time at which the maximum plasma *level* was achieved.

Even if "plasma concentration" and "plasma level" here are true synonyms, using both in the same paper may confuse your readers. Choose one and stick to it. Scientific writing is not literary writing.

The remote verb

One of the most common errors in scientific writing is the use of the "remote verb." In the sentence below, 37 words and numerals separate the subject (children) from the verb (were invited), quoted from a dissertation:

> All children (n = 99, 54 boys and 45 girls) born between 1990 and 1995, adopted during 1993–1997 from Poland, Romania, Russia, Estonia, and Latvia through authorized adoption agencies in Sweden and living in the region of Västra Götaland, were invited to participate in the study.

The separation could have been avoided by beginning the sentence with the first person, active voice, followed by the original last five words:

> We invited to participate in the study all children . . .

3

Drafting the manuscript

As no two authors write in the same way, no one can say which way of writing will suit you best. You will have to find out for yourself. The writing procedure described here is the one I personally have found most useful – by trial and error. Hopefully you may find some portion of it to adopt.

A central part of this writing scheme is to collect ideas while the study is in progress.

Write down your thoughts as they arise

While the study is still in progress, jot down ideas as they occur to you. The notes can be assembled, for example, in a loose-leaf binder containing plastic sleeves, one for each section of the paper. (Woody Allen, the moviemaker, works in a similar way; in a drawer he gathers slips of paper with ideas for his forthcoming movie.)

Ideas can pop up anywhere – in bed, in the bath, in the street, on the bus, on the train, in the car. So, place your notebooks strategically so that you always have one at hand, wherever you are. Use one sheet of paper per idea, even if the idea is only a single line or phrase. Eventually, the reservoir may contain all the components of the paper (or film script), waiting to be arranged.

Where and when to write?

As a beginner I made the cardinal error of taking two weeks off and sitting down on a Monday morning in an attempt to write the first draft continuously from beginning to end. It didn't work. Professional writers don't do it that way. They know from experience that they can work creatively for only a few hours per day. They also know the importance of working uninterruptedly, with no phone ringing and no visitors arriving. For example, when Vilhelm Moberg, author of the great epic about Swedes leaving their homeland for America, was at one time writing in California, he could find only one place to work undisturbed – in the attic of the house. No one could reach him there, because he pulled the ladder up behind him.

Writing an epic is, of course, not the same as writing a scientific paper, whose well-defined sections can be used to divide the text into separate stages. Short sections such as abstract and introduction may be written in a single session each. Long sections such as results and discussion might have to be split into smaller parts, each to be written in one session.

How it can work in practice

Assume that during this particular sitting you intend to write the introduction. You have three hours at your disposal. Before starting, read and revise what you wrote during the previous sitting. Then read the notes you have collected for the introduction. Let us say that writing this section takes only about two of the three hours you have available. Nevertheless, stop writing now – it will give you a feeling of accomplishment. However, before you finish for the day, read the notes you have collected for the next part and sketch the main topics in brief, incomplete sentences.

Even if you still have 20 minutes to spare, and are still full of energy and creativity, do *not* start writing the next part. If you do, you may have to leave the work uncompleted, with a feeling of dissatisfaction. Ernest Hemingway once said about writing, "Always stop on a high," and that is exactly what you do if you always stop when one part is finished.

Medical researchers with clinical duties rarely have as much as three hours of uninterrupted time available. But this writing program can be used for shorter (1–2 hour) spells if you adjust the pieces of work accordingly. A great advantage of this writing scheme is that you need not write every day.

The other way around

You do not have to begin with the abstract or the introduction. You may begin by writing the easiest section, which could be the methods or the results. This approach offers a psychological advance. Starting with the information you know best (the methods or the results) gets about a third of the paper done quickly, and you look forward to writing more. Then, feel free to write the remaining sections in whatever order you find easiest.

Handwriting or word processing?

Handwriting may be suitable for the first draft, but word processing is without doubt the easiest method for revision. If you are going to revise a section extensively, make a copy of the original version and save it in a separate file – you may need it if you change your mind.

4

Choosing a journal

You will most probably find the right journal for your paper among those periodicals you most often read. That is where you have your readership.

If you think that more than one journal seems appropriate, you may wish to rank them by quality. One way to do so is to look at the "impact factor," which tells how often the average article of a journal is cited. Such information is provided by the Institute for Scientific Information in its annual *Journal Citation Reports*.

The impact factor is especially useful for comparing journals within a particular field of research. Let us take, for example, *Orthopaedics*. The 41 journals listed for 2005 had an impact factor in the range 0.1–4.2, with a median of 0.9. It is reasonable to assume that journals with an impact factor of 4.2 attract the best papers in the field, and that these journals have a greater impact on science in that field than a median (0.9) impact-factor journal.

However, if you select a high-impact journal, the publication of your paper may be delayed, as is hinted at in this question from a course participant:

> Should I send my paper to a journal with a high impact factor
> and risk having it rejected, or should I send it to a journal with
> a lower impact factor and get it published quicker?

How to Write and Illustrate a Scientific Paper

If you feel in your heart that yours is a first-class paper, then try the high-impact journal – provided that it is a specialized journal in your own field. However, if it is a journal outside your specialty and your paper is accepted and published, this journal might turn out to be a publication that researchers in your specialty do not read. For example, a colleague of mine complained that his excellent paper published in one of the highest-ranked medical journals, *The Lancet*, was not cited. However, when you have been around for a while, you may feel by intuition which journal is the right one.

The impact factor ranks journals; it does not evaluate individual papers. Some articles may not be cited at all, while others become classics. Although it may be outside the scope of this book, I will tell you about one way to find the best papers in your field: visit the website Faculty of 1000 Medicine (www.f1000medicine.com) or Faculty of 1000 Biology (www.f1000biology.com). These sites rate individual papers according to their merit, irrespective of where they are published. However, you must be aware that a top-ranked article may not necessarily be well written.

Instructions to authors

When you have chosen a journal, the next step is to read the current version of its Instructions to Authors. Several journals print these instructions in every issue, others only in the first issue of each volume. They also appear on the website of the journal. If you work in a biomedical discipline you will find that many journals use "Uniform requirements for manuscripts submitted to biomedical journals" (Vancouver Document, www.icmje.org), a set of instructions intended to allow authors to use the same format and style for papers submitted to different journals.

5

Preparing a graph

Assuming that your results show trends or movements over time, such as nicotine concentration in plasma after smoking, a good way to display your data would be to construct a line graph. But do not rely on the computer to design it for you. Here are some common errors.

The line graph

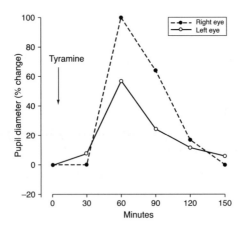

Figure 5.1 The effect of tyramine solution on pupillary size. (Adapted, with permission, from a draft by Havelius 1994.)

How to Write and Illustrate a Scientific Paper

This seemingly excellent line graph nevertheless appears to have two common defects: the curves are distinguished both by type of line and by type of data-point symbol – either would suffice; and the curves are identified by a separate key, obliging the reader to scan back and forth to the key to see what they represent.

In the two redrawn graphs (5.2), the curves are labeled directly and distinguished either by type of data-point symbol or by type of line.

Open and filled circles, as in the left graph, are the data-point symbols easiest to distinguish. They can also be used symbolically; for instance, if an experiment has been performed with (●) and without (○) treatment, the emptiness of the open circle suggests that nothing has been administered.

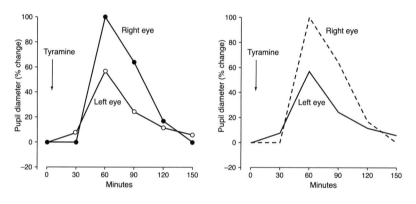

Figure 5.2 Alternative displays of Figure 5.1

Other standard symbols for data points are open and filled squares and triangles (□ △ ■ ▲). If you need more symbols, you probably have too many curves for one graph, and you should consider dividing it into two or presenting your observations in a table.

In the right-hand graph you will probably not miss the data points, as you can easily discern the change of line direction where the points have been omitted. This graph may be the more attractive of the two. Data points are probably overused in scientific papers.

Relationship between the lengths of the axes

In the following graph (5.3), the sharp decrease in the first part of the line has been exaggerated in two ways: (1) the vertical axis is longer than the horizontal axis and (2) the horizontal axis is contracted because the distance between the first two tick marks represents four hours whereas the same distance between the following ticks represents only one hour.

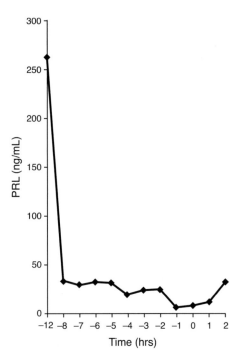

Figure 5.3 The original line graph exaggerating the decrease. (Reproduced with permission from *Acta Obstetricia et Gynecologica Scandinavica* 2001;**80(1)**:34–8.)

The golden ratio, which is close to the format 3:2, is the most aesthetically pleasing. But the 3:2 may invite misuse. Therefore, the relationship between the axes should normally be 1:1, as in the redrawn figure (5.4) on the next page.

Time point zero represents the time of delivery; this information is given in the main text, but it could have been included in the figure.

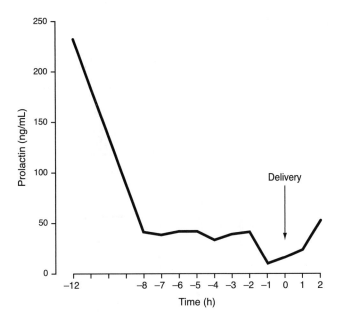

Figure 5.4 Recommended version of Figure 5.3

Labeling axes

The reader must be in no doubt as to what the axes show. A few drawing programs do not label the vertical axis parallel to the axis but instead place a line of text above the figure, like a title. I have even seen this practice used to describe the horizontal axis! So always place the label parallel to the axis.

A graph is intended to show a trend, not exact figures. Therefore, the number of tick marks should be limited. On the vertical axis, the "*1–2–5*-principle" is often used; that is, the axes are divided into intervals of *1*, 2, 3, …; *2*, 4, 6, …; or *5*, 10, 15 …; and not 7, 14, 21, etc. You may, if necessary, multiply by 10 (Figure 5.4) or 100, but preferably not by 1000. As scientific units are often expressed in multiples of thousands, three zeros are easily removed by altering the unit, for example, from microgram to milligram.

The chart

A time series can also be displayed as a set of vertically arranged bars, known as a *column chart* in most computer programs. Here is an example.

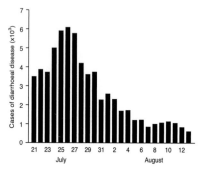

Figure 5.5 Reported cases of diarrhoeal disease (cholera, dysentery, and dehydration). (Reproduced, with permission, from Goma Epidemiology Group, Public health impact of Rwandan refugee crisis: What happened in Goma, Zaire, in July 1994? *The Lancet* 1995; **345(8946)**: 339–44, © The Lancet Ltd.)

This brings us to the question: Which type of presentation is preferable for a time series such as this – column chart or line chart? As a basis for discussion I have redrawn this time series as a line graph (5.6). To avoid ambiguity, I have converted $x10^3$ to *thousands* on the vertical axis. Note that the curve is bolder than the axis lines. Note also that the axes are separated, as zero is not common to both.

Now let us compare the two designs, step by step. A column chart may be preferable when there is no carry-over effect from one time period to the next, that is, when each column represents a new set of data with no addition from the preceding time period. For example, the annual number of births could be shown as a series of columns, while the total population could be plotted as a continuous line (Chapman and Mahon 1986). As there is no carry-over effect in the example of diarrheal disease, the lines joining the data points are, strictly speaking, artifacts. Those who worry about such things might prefer a column chart.

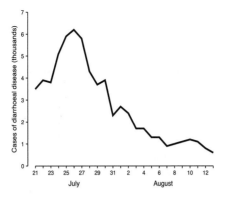

Figure 5.6 Alternative display of Figure 5.5

Column charts are said in certain cases to exaggerate differences between individual measurements. If this is so, it could be a reason for not using column charts in such cases.

A figure should have no unnecessary lines. The column chart (5.5) has 72 lines (three for each of the 24 columns), whereas the curve graph has only one line. In a line graph the viewer can usually see trends or movements more quickly than in a column chart. Which graph would you have chosen in this case? I would probably have preferred the line graph.

One type of illustration most readers dislike is those *grouped column charts* that have more than two or three categories in each group, as in the following example (5.7).

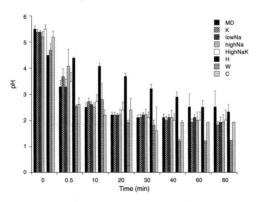

Figure 5.7 Beverage and gastric contents pH. (Reproduced with permission from *Medicine and Science in Sports and Exercise* 1993; **251(1)**: 42–51.)

The key to this figure needs a key of its own. *W*, for example, appears not to stand for *watt*, as it generally does, but *water*. *C* doesn't stand for *carbon*, but *controls*. Suppose you now look away from the columns. What do you remember having seen, without looking back? This type of data might well do better in a table.

Two or three categories in each group should be the maximum in the grouped column chart. Figure 5.8 with two categories is easy to grasp.

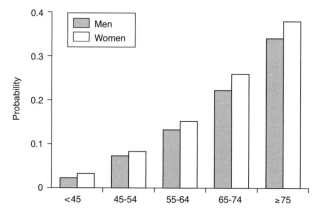

Figure 5.8 Probability of dying in a coronary care unit after admission with initial working diagnosis of acute myocardial infarction. (Reproduced, with permission, from Clarke *et al.* 1994.)

One way to remove the separate key in this graph could be to label the first group directly, for example:

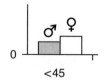

However, the main role for column charts is *not* to show time series (such series are usually better shown by line graphs), but to display *categorical data*. The following graph (5.9) presents such data. What makes this graph especially good is that it shows the precise percentage for each of the columns. Such a design combines the virtue of the table (giving exact values) with that of the graph (quickly presenting the message). Note that the bars are wider than the spaces between them. Note also that the bars have a gray tone (somewhere in the middle of the gray scale), which is more pleasing to the eye than a black or white tone or a striped pattern.

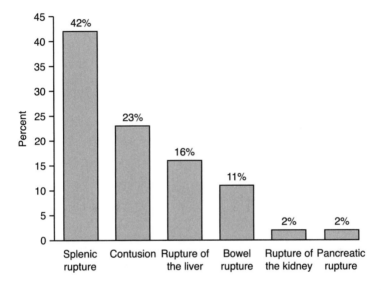

Figure 5.9 Intra-abdominal injuries associated with diaphragmatic rupture due to blunt trauma. (Reproduced, with permission, from Sarna and Kivioja 1995.)

However, the cramped space of a journal column (about 8 cm) allows a column chart to include only a few items. This chart is on the verge of being overcrowded. Below the two columns on the far right, the two separate texts nearly run into each other: *Rupture of Pancreatic . . .* Such a problem can be overcome with the use of a *bar chart,* the computer term for horizontally arranged bars.

The next figure (5.10) is a good example of a bar chart. The bars are arranged in decreasing order of size. The chart also shows the exact numerical value of each bar.

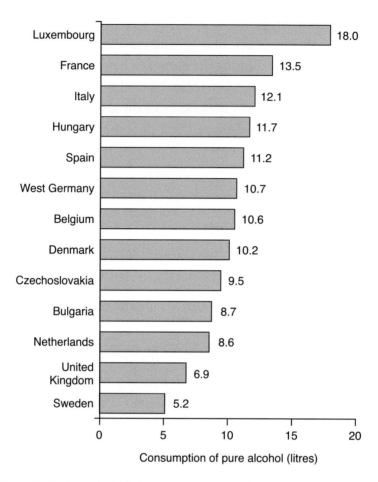

Figure 5.10 Annual alcohol consumption per inhabitant. (Reproduced, with permission, from van Os and Neeleman 1994.)

We will turn to quite another item: the presentation of *individual* data. I will show how much more information you can give the reader by displaying raw data instead of summary data.

The following graph (Figure 5.11) shows individual data and can be regarded as an opened-up column chart. At the bottom of the right column, seven outliers (outside values) attract attention (arrow mine).

How to Write and Illustrate a Scientific Paper

The seven outliers turned out to have an atypical variant of Fabry's disease, a serious metabolic abnormality. Had a summary-data graph been used instead (with columns having almost the same mean and standard deviation), the outliers would have been effectively camouflaged.

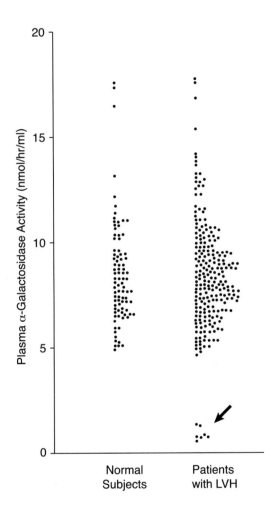

Figure 5.11 Plasma α-galactosidase activity in 89 normal male subjects and 230 male patients with left ventricular hypertrophy (LVH). (Reproduced, with permission, from Nakao *et al.*, An atypical variant of Fabry's disease in men with left ventricular hypertrophy, *N. Engl. J. Med.* 1995; **333**:288–93. Copyright © 1995 Massachusetts Medical Society. All rights reserved.)

Now we will do the opposite. We will peek behind the columns of a summary-data chart (5.12) at the raw data to find out what they can tell us. The chart I have chosen shows the time needed to induce abortion in the second trimester when using a conventional method (prostaglandin alone) compared with a new approach (prostaglandin + oxytocin).

Clinicians in the field seeing this graph may at first be impressed by the seemingly dramatic shortening of the abortion time. However, they would probably be less impressed when viewing the redesign (Figure 5.13) consisting of individual data from the 53 subjects;

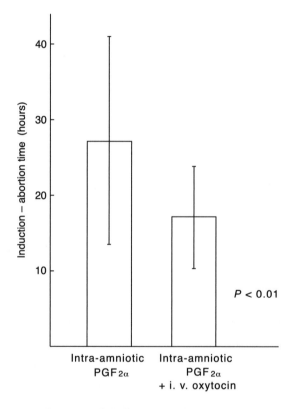

Figure 5.12 Interval (mean and SD) from induction to abortion for women treated with intra-amniotic prostaglandin F2α (PGF2α) alone or in combination with intravenous oxytocin. (Reprinted from *Prostaglandins*, 2, M. Seppälä, P. Kajanoja, O. Widholm, P. Vara, Prostaglandin-oxytocin abortion: A clinical trial on intra-amniotic prostaglandin F2α in combination with intravenous oxytocin, 311–9, 1972, with permission from Elsevier Science.)

How to Write and Illustrate a Scientific Paper

data obtained from two spike bars (also called needle bars; Harris 2006) in the same paper.

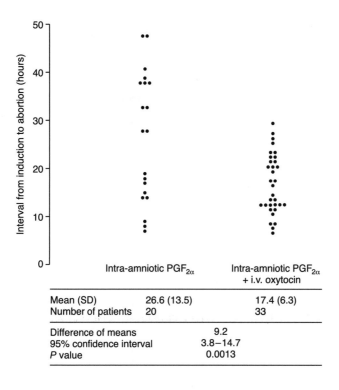

	Intra-amniotic PGF$_{2\alpha}$	Intra-amniotic PGF$_{2\alpha}$ + i.v. oxytocin
Mean (SD)	26.6 (13.5)	17.4 (6.3)
Number of patients	20	33
Difference of means	9.2	
95% confidence interval	3.8−14.7	
P value	0.0013	

Figure 5.13 Alternative display of Figure 5.12

The redrawn graph shows that the improvement was not as clear-cut as might be expected from the summary-data chart. I also combined the graph with a table containing complementary data, so that the reader does not need to search in the main text for this information – a design recommended by Altman (1995, 40).

Altman (1995, 222) and Tufte (1983, 13–14) show with illustrative examples how the *same* summary data can rest on entirely different sets of raw data. So before you start writing your paper I recommend that you make, at least in sketched form, a graph with your individual observations plotted, to see what you might find.

If you eventually decide to present your results as summary data, avoid the following type of display.

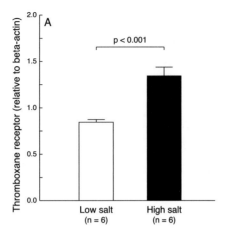

Figure 5.14 Data (mean ± SE) compare rats adapted to low-salt with those adapted to high-salt intake. A, kidney cortex. (Reproduced from Welch *et al.* 1997, with permission from the American Physiological Society.)

To indicate a single mean value with a column is redundant. In the following redesign the columns have been removed and the means shown with data points.

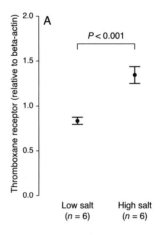

Figure 5.15 Alternative display of Figure 5.14

How to Write and Illustrate a Scientific Paper

Here are some further comments on the original figure (5.14). There is no need to distinguish the bars by making one of them white and the other black; both bars in a gray tone would have been preferable (see Figure 5.9). Note also that the label of the vertical axis is too long, extending beyond the axis. An asset of the graph, however, is that it presents the level of probability with a P value rather than an asterisk. The use of asterisks (*$P < 0.05$, **$P < 0.01$, ***$P < 0.001$) should be avoided; instead, give exact P values (the exception is $P < 0.001$; see Chapter 22, "$P < 0.05 \neq$ the truth").

The *box-and-whisker plot (box plot)* has become a popular form of presentation of data. As there are many variations (Harris 2004), you will have to explain the details of the plot, as in the following example from *The Lancet* (Chaparro *et al.* 2006):

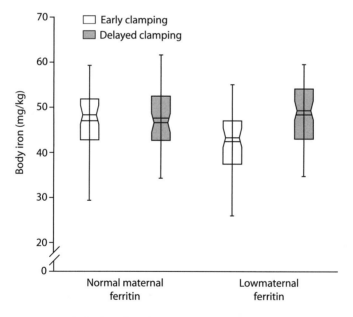

Figure 5.16 Box-and-whisker plot of two-way interaction effect of treatment group and maternal ferritin on infant body iron (mg/kg) at 6 months of age
Boxes represent the inter-quartile range (25th to 75th percentile), and whiskers indicate the 5th and 95th percentiles for unadjusted data. The notch in each box represents CI about the median, represented by horizontal line at the middle of the notch. Additional horizontal line represents the mean of each subgroup. [...] (Reproduced from *The Lancet*, **367**, Chaparro *et al.*, Effect of timing of umbilical cord clamping on iron status in Mexican infants: a randomised controlled trial, 1981–9, © 2006, with permission from Elsevier.)

If you are uncertain whether to present your results as summary data or as individual values, you can use both presentations, as in Figure 5.17.

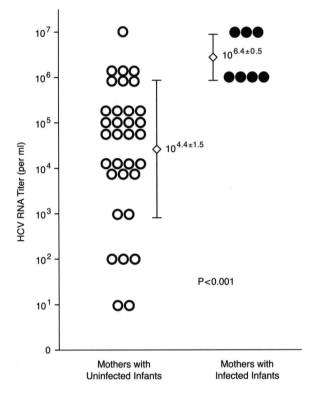

Figure 5.17 Mean (±SD) serum HCV [hepatitis C virus] RNA titers in the 33 mothers with uninfected infants and the 7 mothers with HCV-infected infants. (Reproduced, with permission, from H. Ohto, S. Terazawa, S. Sasaki, N. Sasaki, K. Hino, C. Ishiwata *et al.* Transmission of hepatitis C virus from mothers to infants. *N. Engl. J. Med.* 1994; **330**:744–50. Copyright © 1994 Massachusetts Medical Society. All rights reserved.)

Another way to present both individual and summary data is to combine an individual-data graph with a table, as in Figure 5.13. One can wonder why these two informative ways of presentation (Figures 5.13 and 5.17) are so seldom used.

Let us now discuss the three-dimensional graph. Today's computer technology renders the three-dimensional graph easily displayed. As a result, this type of graph is seen more and more

How to Write and Illustrate a Scientific Paper

often in published reports. Unfortunately, the ease with which it can be created often leads to its use even in cases where the data have only two dimensions. A third dimension is thus falsely introduced in such cases. Figure 5.18 is a typical example.

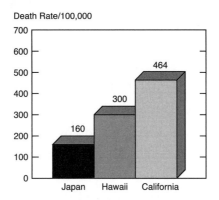

Figure 5.18 Mortality from coronary disease among Japanese-born men 55 to 64 years of age residing in Japan, Hawaii, or California, 1950. (Reproduced, with permission, from Reed 1990.)

After conversion into a two-dimensional display, the graph is easier to read (Figure 5.19).

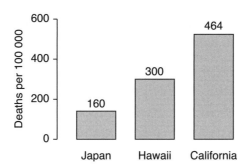

Figure 5.19 A two-dimensional display of Figure 5.18.

Note that the bars have been shifted apart and given the same gray tone. The bars are wider than the spaces between them. Note also that the vertical axis does not extend beyond what the graph demands, numbers on the scale are reduced, and tick marks point outward.

A *true* third dimension is extremely uncommon in research results. However, Tufte presented an excellent example of this rare species in his classic *The Visual Display of Quantitative Information* (1983, 42), where he depicted air pollution over six counties in southern California. Only thanks to the three-dimensional presentation can one distinguish between the peak over San Bernardino, in the background, and that over Los Angeles.

Pie charts may be of two kinds: three-dimensional, called *pie charts*, or two-dimensional, called *simple pie charts*.

The three-dimensional *pie chart* is less suitable for presentation of scientific data, as it is difficult to compare three-dimensional segments (especially between charts of different sizes) and because of the waste of space (few data per area of graphic). Tufte (1983, 178) says that "the only worse design than a pie chart is several of them."

The *simple pie chart,* however, is used increasingly in scientific papers. As it is more illustrative than scientific, its usage may be appropriate in a magazine article. Although I am not a supporter of its employment in scientific writing, I will nevertheless describe how to utilize it. A good simple pie chart has four characteristics (Figure 5.20): (1) the largest segment begins at 12 o'clock; (2) it continues with proportionally smaller portions in the clockwise direction; (3) the number of segments does not exceed five; and (4) labels are placed outside the circle. For emphasis, one sector can be separated slightly; most software programs allow you to do this. If, however, the space is cramped, the content of this figure (5.20) could be given in the running text as follows:

> Of the 20 patients studied, 12 had myoma uteri; 6, adenomyosis; and 2, endometriosis.

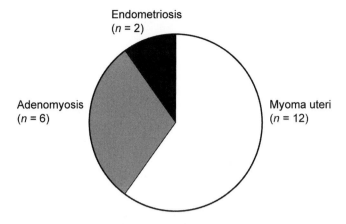

Figure 5.20 Clinical diagnosis of the patients studied. (Adapted, with permission, from a draft by Chyi-Long Lee and Yung-Kuei Soong 1991.)

Before submission

Any graph must withstand reduction to a journal column width (about 8 cm). After reduction, the text of the axis labels should be similar in size to the running text. Reduce your graph on a photocopier to see if that will be the case.

When manuscripts are submitted electronically, illustrations may either be included with the text or uploaded separately. Consults the Instructions to Authors of the journal you have chosen to find what they prefer.

6

Drawings

Although my first drawing could have been published with only slight improvements, I let an artist redraw it, by hand. (We had no computers in those days!) The picture was improved – but it cost me a small fortune. Ever since, I have done my drawings myself. This is perhaps not such a bad thing after all, because the pictures will express exactly what I mean.

Such a drawing, done by one of the authors of a paper, is shown below.

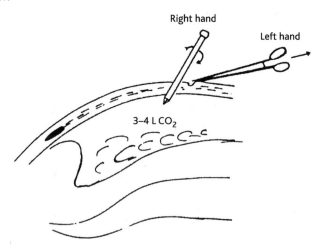

Figure 6.1 How to prevent injuries to internal organs when inserting the trocar before performing a laparoscopy. (My translation from the Swedish. Illustration adapted, with permission, from Samuelsson and Sjövall 1973.)

The report was initiated by a serious complication to a laparoscopy, where a trocar (a sharp instrument) was stabbed into the large intestine during the initial phase of the procedure. The paper describes how this complication could be prevented – and the drawing tells exactly what the authors wanted to say. The picture cost them nothing!

Authors who wish to have their drawings done by a professional artist are recommended to provide the artist with a detailed sketch. The artist then scans the sketch into a computer where it is improved – much cheaper than drawing it by hand. Skilful illustrators, however, are a rare species and usually have a tight schedule. You should therefore contact them in good time; preferably before you begin to write the paper.

The following pair of figures shows the sketch (above) done by the author and the improved drawing (below) by the artist. Every detail is enhanced and given a professional look.

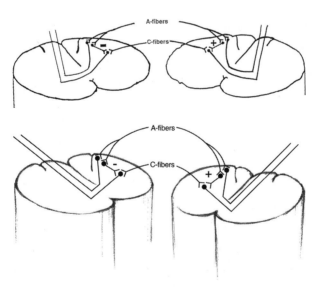

Figure 6.2 Drawing in sketch form (above) and final form (below). (The sketch form printed with the permission of the author, Joanna Wallengren. The final form reprinted from *Journal of the American Academy of Dermatology*, **39**, Wallengren, J. Brachioradial pruritus: A recurrent solar dermopathy. 803–6, Copyright (1998), with permission from The American Academy of Dermatology, Inc. Illustration by Ronny Lingstam.)

7

Figure legends

A figure legend or caption may include two items:

(1) the **title**, which states the topic of the figure, and

(2) the **message**, which explains the contents of the figure.

In some cases a message alone will suffice.

However, most legends do not convey a message. Figure 7.1, with its typical legend, could be improved as shown in Figure 7.2.

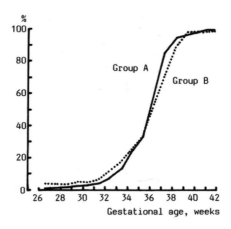

Figure 7.1 Cumulative weeks to delivery of women in group A (n = 78) and group B (n = 78). (Reproduced from a manuscript in preparation that after revision was published in *Acta Obstetricia et Gynecologica Scandinavica* 1988; **67**:81–4; with permission from the author and Munksgaard International Publishers Ltd., Copenhagen, Denmark.)

In the revised illustration, *Group A* has been changed to *Treated women* and *Group B* to *Controls* so that the reader does not have to search in the main text to interpret these terms. The tick marks now face outwards, so that they do not encroach on the curve between 26 and 30 on the horizontal axis.

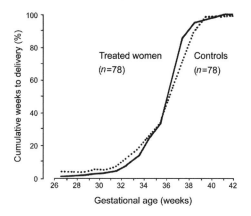

Figure 7.2 Outcome of twin pregnancy in women prescribed either prophylactic leave of absence from work (treated) or not (controls). Gestational duration did not differ between the groups.

An illustrative example of a legend including the message alone is presented in Figure 8.1.

8

How to design tables

The content of a table may be either descriptive, as is most often the case, or declarative (bearing a message). It helps the reader if this is reflected in the table's title.

The descriptive title

The descriptive title is used for tables that present detailed information, such as the one below (only part of the original table is shown).

Table 8.1 Maternal age, gestational age, indication, size and type (avascular or vascular) of villi sampled, method used in processing the biopsy (direct preparation, 24 h culture, long-term culture), and karyotype in 80 diagnostic cases of first-trimester chorionic biopsy

| | | | | Villi sampled | | | Cytogenic method | | | |
Case No.	Maternal age (y)	Gesta-tional age (wk)	Indication	Weight (mg)	Avas-cular	Vas-cular	Direct	24 h culture	Long term culture	Karyotype
1	45	12	Previous child Mb Down	10		X			X	46,XY
2	19	12	Hemophilia	2					X	46,XY
3	30	9	Hemophilia	10		X			X	92,XXYY
				5	X					
4	41	11	Age	8		X			X	46,XY
[etc.]										

Source: Reproduced from Heim *et al.* 1985, with permission from Munksgaard International Publishers Ltd., Copenhagen, Denmark (partial table).

This table is from a paper of which I was a coauthor. Now, with the benefit of hindsight, I see that this title contains the common error of repeating every single heading, except, in this case, the first one. No fewer than 31 of the 39 words in the title can be deleted. The shortened version can be grasped at a glance:

Table 8.1 Details of 80 diagnostic cases of first-trimester chorionic biopsy

The declarative title

If a table shows a clear trend or relationship, a declarative title could be preferable. The following table shows that the frequency of fractured clavicles increased with increasing birth weight.

Table 8.2 Fractured clavicles and birth weight   ← Title

Birth weight, g	Deliveries		Fractured clavicles	
	n	%	n	%
–2500	434	8.5	9	2.07
2501–3000	1395	27.3	45	3.23
3001–3500	2047	40.0	108	5.28
3501–4000	1049	20.5	111	10.58
4001–	193	3.7	24	12.44
All	5118	100.0	297	5.80

← Headings

← Body

Source: Reproduced from a manuscript by Jójárt et al. 1992, with permission.

But the title is neutral and one has to study the table for a while in order to grasp the message. It would have helped the reader if the table's message had been stated in the title:

Table 8.2 Increase in fractured clavicles with birth weight

Rounding off

In the table's far right-hand column, the percentages shown are over-precise. One decimal would suffice (see Chapter 20, "Percentages"). In the middle column, the percentages add up to exactly 100.0 percent. The authors have achieved this by adjusting the true percentage of 3.8 to 3.7 (second figure from bottom). *Do not on any account adjust figures to make them add up to the true total.* Instead, give the total after rounding (in this case 100.1 percent) and explain in a footnote, "The sum of the percentages exceeds 100 percent due to rounding."

Having said this, I also want to emphasize the assets of this table. What is especially good about it is that it makes only one single point. If, on the other hand, a table's data lead to two conclusions, the author had better try to divide the table into two smaller ones, as two conclusions in the same table tend to obscure each other. Another asset of this table is that numbers that are to be compared follow down the columns, not across. They are easier to read that way. (If you don't believe me, try adding percentages from left to right in a row!)

Table or graph?

As mentioned earlier, a table can be used either to report precise numbers or to illustrate a trend. But often a trend is better illustrated with a figure. Here is an example showing what I mean.

In its day, the following table (8.3) contained medical dynamite. It presented the results of a study that showed, for the first time in vivo, that certain anti-inflammatory substances, such as indo-methacin or aspirin, inhibit the synthesis of prostaglandin. The sensational outcome of this study is well hidden among the figures within the table. Moreover, the title reveals nothing. And to open a title with a long chemical name is deadening; *Prostaglandin metabolite* would have sufficed; the full chemical name belongs to the methods section.

Table 8.3 Excretion of 7α-hydroxy-5, 11-diketotetranor-prostane-1, 16-diodic acid in subjects receiving analgesics. Indomethacin (a, 4x50 mg/24 h [. . .] given as indicated by asterisk.

Subject	Amount of metabolite (mg/24 h)						
	Day 1	Day 2	Day 3	Day 4	Day 5	Day 6	Day 7
Ia	4.8	4.8	1.8*	1.1*	1.5*	2.7	4.1
IIa	3.9	4.4	0.7*	0.7*	0.7*	3.1	6.5
IIIa	3.8	3.0	0.5*	0.3*	0.3*	0.8	1.1
[etc.]							

Source: Reproduced, with permission, from Hamberg 1972 (partial table).

Shown here (Table 8.3) is only the part of the table that includes the results from the three subjects receiving indomethacin. I converted this part into a graph (Figure 8.1) to show more clearly the dramatic effect of indomethacin.

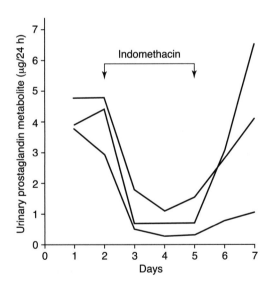

Figure 8.1 Urinary excretion of a prostaglandin metabolite decreased following indomethacin administration in three humans

Note that the curves have the same type of line because they do not need to be distinguished from each other; they are all intended to show the same trend. The curves are bolder than the axes. Two zeros are used to label the point where the axes meet. Tick marks point outward. Axes are of equal length. The label of the vertical axis is parallel to the axis and reads from bottom to top. Note also that the legend gives the message of the figure.

On the next page, I will show how to design a table. To present all these details in the running text would bore you. Instead, I have arranged the instructions in note form around an imaginary table. With the exception of the footnotes, the text of the table is *Once upon a time*, typed repeatedly. So, you can concentrate on the layout without being distracted by the content of the text.

How to Write and Illustrate a Scientific Paper

The Vancouver Document (2002) recommends that footnotes be indicated with superscripts *, †, ‡, §, ||, ¶, **, ††, and so on – which is one of the few recommendations of the Document I disagree with. In this book, I have used the more plain sequence a, b, c, and so on, which is seen increasingly in biomedical publications.

If serveral abbreviations or symbols are used in a table, all can be collected and defined in a general footnote keyed to the table title (CBE 1994, 686).

Only three full-width lines; no vertical lines.

Table 8.4 Once upon a time[a]

	Once upon a time		
	Once	Upon	A time
Once	(xx)	(xx)	(xx)
Once	xxx	xxx	—[b]
Once upon[c]	xxx	xxx	xxx
Once upon a time once	xxx	xxx	xxx

[a]Data are given as means (SD) unless stated otherwise. Hb denotes hemoglobin, RBC red blood cell count, and WBC white blood cell count.

[b]No measurement made.

[c]Median (interquartile range).

Runover lines indented (at least 5 mm) and single-spaced (the only exception from double spacing in a manuscript).

Under the first table in an article, all abbreviations are explained but need not be repeated under subsequient tables, where this footnote suffices: "Abbreviations as in Table 1."

9

Title

For every person who reads the whole of a scientific paper, about 500 read only the title (Kerkut 1983). One way to improve this statistic could be to make the title declarative by including what the paper says, not just what it covers.

Whenever possible, use a declarative rather than a neutral title

This title is neutral:

> Influence of aspirin on human megakaryocyte prostaglandin synthesis

John Vane, in his classic paper published in *Nature* in 1971, put it more expressively:

> Inhibition of prostaglandin synthesis as a mechanism of action of aspirin-like drugs

(In 1988, Vane was awarded a Nobel Prize in acknowledgment of his discovery. Vane told us how aspirin relieves pain.)

How to Write and Illustrate a Scientific Paper

The following declarative title is taken from the biological sciences (Marvin 1964):

> Birds on the rise

Goodman *et al.* (2001) recommended that the study design also be included in the title, as follows (Lee et *al.* 1995):

> Improved survival in homozygous sickle cell disease: Lessons from a cohort study

However, the following formulation is *not* recommended (Quesada *et al.* 1995):

> Leaf damage decreases pollen production and hinders pollen performance in Cucurbita texana

Do the authors really mean to imply that the issue is settled once and for all? Your own present work should be referred to in the *past* tense:

> Leaf damage decreased . . . and hindered . . .

This way of using a verb in a title makes it into a sentence. It is stronger than using a phrase, some say too strong. In a descriptive study, however, you have to use a neutral title (Kitin *et al.* 2004):

> Anatomy of the vessel network within and between the tree rings of Fraxinus lanuginosa (Oleaceae)

Titles ending with a question mark

Scientists used to have the journal at hand when skimming through the table of contents and could page straight to the article. Today, most scientists skim lists of titles on a computer screen but frequently are unable to get access to the article and sometimes even the abstract. Thus, instead of a title in form of a question like this (McWhorter and Martínez del Rio 2000):

> Does gut function limit hummingbird food intake?

the reader would appreciate being told the answer from the beginning:

> Limitation in hummingbird food intake by gut function

A review article can, however, have a title ending with a question mark, because some papers reviewed say one thing; others, the opposite. The title of a mini-review published in *Nature* (Pitnick *et al.* 1995) presumably covering all or most facets of the debate is, for example:

> How long is a giant sperm?

Begin with the keywords

Not until the final words does the following title tell which disease was being studied. Thus a reader who is in a hurry when scanning the table of contents may not have time to find out what it is about.

> The effect of calcium antagonist felodipine on blood pressure, heart rate, working capacity, plasma renin activity, plasma angiotensin II, urinary catecholamines and aldosterone in patients with essential hypertension

When the keywords are placed at the beginning of the title, it is immediately clear what disease was studied:

> Essential hypertension: The effect of . . .

Use verbs instead of abstract nouns

The following is a common way of formulating a title:

> Treatment of polycystic ovary syndrome

but if you turn the noun into a verb, you will make the sentence more dynamic:

How to treat . . .

Avoid abbreviations in the title

In *The Lancet* (1993), which asks us to avoid abbreviations in titles, I saw the following heading for an editorial (the editor has denied that it was a joke):

OCs o-t-c?

Written out in full it means "Oral contraceptives over-the-counter?" or, in other words, should pharmacies be allowed to sell oral contraceptives without a doctor's prescription?

Admittedly, I read this editorial because its unintelligible title caught my attention. So this instance could be regarded as an exception from the rule of thumb, not to use abbreviations in the title.

Of course, you are allowed to use abbreviations and symbols that are more familiar than the words they stand for, such as DNA and pH. In case of doubt, however, use both the term in full and its abbreviation, as in this title (Stockdale 2000):

Contaminated material caused Creutzfeldt-Jacob disease (CJD) in some undersized children who were treated with growth hormone (GH)

A title for your thesis

In *JAMA* I saw the following title, which is composed of a main heading for the broad aspect and a subheading for the details (Hodgen 1981).

Antenatal diagnosis and treatment of fetal skeletal malformations
with emphasis on in utero surgery for neural tube defects and limb bud regeneration

This type of title could be useful for your doctoral dissertation: the *main heading* for the nonspecialists, including your relatives, neighbors, and friends; and the *specific subheading* for experts in the field.

Running title

As an aid to readers, most journals print a running title at the top (running head) or bottom (running foot) of every page or alternate pages. Thus, if a journal is opened in the middle of an article, the reader will know what it is about.

As your main title will probably be longer than the stated limit for the running title, provide (on the title page of the manuscript) an abbreviated version. When doing so, focus on the keywords, as did the authors of this title of a paper published in *Obstetrics & Gynecology* (López-Jaramillo *et al.* 1997):

> Calcium supplementation and the risk of preeclampsia in Ecuadorian pregnant teenagers

which was condensed to:

> Calcium supplementation reduces preeclampsia

However, as observed earlier, the presentation of results in the current paper should be in the *past* tense, thus "reduc**ed**"; *present* tense, "reduc**es**," is used for established knowledge.

A course participant, listening to my advice regarding these principles, asked, "Do you really mean that Tjio and Levan [who discovered that we have 46 chromosomes, not 48] should have entitled their paper *The chromosome number of man **was** 46*?" Of course not, that would have misled the reader completely. This is an illustrative example of Rule No. 1 of writing a scientific paper: the author's common sense should always take precedence over the principles of authorities.

In the running title discussed here, however, the *past* tense "reduc**ed**" is acceptable.

Calcium reduced the risk of preeclampsia

Note that I have reinstated "the risk of," because a condition cannot be reduced – only its frequency, degree of difficulty, or both can.

Hormone replacement therapy and risk of non-fatal stroke
Anette Tønnes-Pedersen, Øjvind Lidegaard, Svend Kreiner,
Bent Ottesen

and here is part of the contributors list:

> Anette Tønnes Pedersen carried out the study, as part of her
> PhD project, and was responsible for all parts of the research
> project, including the writing of the paper. Øjvind Lidegaard
> was responsible for the initial study design . . . Svend Kreiner
> did the statistical analysis . . . Bent Ottesen was responsible for
> overall supervision . . .

Thus, Tønnes-Pedersen did all that could be expected of a graduate student and, in the contributors section, she had the opportunity to say so.

Readers will also be served by such a list. They will find it easier to identify which person to contact to discuss some point in the paper or to request a reagent.

Even if your target journal has not yet adopted the "contributor" concept, you are encouraged to include a contributors list in your manuscript. The editor would love it – and it would also give impetus to the idea.

But don't be too detailed. In *The Lancet* (1997, **350**:620–3) I saw the following statement in the contributors list of a report concerning a survey using questionnaires (family name is fictitious):

> . . . Isabel Moe . . . mailed the letters and questionnaries. . . .

A question of coauthorship

Here is a question from a course participant concerning coauthorship:

> A multidisciplinary study included, among many other things,
> a psychiatric evaluation of groups of patients, according to
> a standardized method. The psychiatrist was coauthor of a
> number of articles. Should he be listed as a coauthor also in all
> future papers based on further analyses of the material – even if
> he in no way takes part in the production?

I do not have the answer. But let us consider the problem from both sides, the psychiatrist's and your own.

Without the psychiatrist's professional evaluation of the patients, which apparently was not routine work, you would have had no material to analyze and no results to report. Maybe you should list him as a coauthor and also ask him to check the manuscripts. He might be able to contribute valuable ideas on parts of the papers dealing with his specialty.

However, you honestly feel that the psychiatrist has already received his due share of the proceeds. You also believe that you no longer require his expertise. If so, in future papers, you might thank him in the acknowledgments. This will also mean that the psychiatrist, too, can write a paper based on this material as seen from a psychiatrist's point of view – and in return mention you in his acknowledgments.

A third way might be that you talk the situation over with the psychiatrist. But apparently this approach is blocked; otherwise you would have used it already.

This discussion shows how tricky it can be to decide on authorship.

11

Abstract

By writing a *working* abstract at an early stage, you will provide yourself with a framework for the rest of the article. It will guide you in deciding what to include and what to omit, thus sparing you tedious rewriting. Then, when you have completed the paper, you can return to the working abstract and recast it in its final form.

Let us consider here *what* to include. In both types of abstract, the conventional (unstructured) and the formally structured, the same four basic sections are required: *Background* (including the purpose of the study), *Methods, Results,* and *Conclusions*.

The conventional abstract

In the conventional abstract, the four basic components are not identified with explicit headings, which may explain why some authors miss out vital parts. In the worst instance, the abstract contains neither background nor conclusion. Many abstracts written in the conventional manner are uninformative and cause editors much unnecessary editorial trouble. The following abstract, quoted from an article in *Contraception*, commences directly with the method:

How to Write and Illustrate a Scientific Paper

ABSTRACT

80 female Wistar rats were employed in this work. They were homogeneously divided into four groups. . . .

Most journals reporting laboratory studies use the conventional abstract, and it is usually applied to Case Reports and Brief Communications. Clinical journals, however, often require a structured abstract for full-length papers.

The structured abstract

The structured abstract differs from the conventional abstract by having a heading for each section. Here is such an abstract of the classic *The chromosome number of man* by Tjio and Levan (1956). The paper was published before abstract sections had come into common use. So, based on the content of that paper, I have composed a structured abstract in order to show what could be included.

> **Background.** It is generally accepted that the chromosome number in humans is 48. But to count chromosomes has been difficult, as they clump and partially cover each other. In this study, cultured cells were treated with solutions that spread the chromosomes and made them easier to count.
>
> **Methods.** Cultured cells from human embryonic lung were treated with both colchicine and hypotonic solution.
>
> **Results.** Among 265 mitoses counted, all but 4 had a chromosome number of 46.
>
> **Conclusion.** The results suggest that the chromosome number in humans is 46, not 48.

In several clinical journals the structured abstract is more detailed, with the methods section split into, for example, *Design, Setting, Patients, Intervention*, and *Measurement*. The Instructions for Authors will tell you what headings to use, if your target journal requires a structured abstract.

The structured abstract has been criticized. It is longer than the abstract in standard format. Its imposed style may be a strait-jacket constraining the author, inhibiting creativity. Its rigid uniformity may bore the reader.

Admittedly, these objections are serious, but the advantages of the structured abstract outweigh its greater length and the extra effort needed for its preparation. In fact, the structured abstract conveys information so accurately and efficiently that readers could be dissuaded from going on to read the rest of the article. And that is not the intention.

Finally, if you cannot avoid an abbreviation in the abstract, you must explain it, the reason being that the abstract will stand alone in abstracting publications. For the same reason, if you are convinced that your abstract must include a reference to a significant work, give a full reference.

12

Introduction

Michael Crichton, the author of *Jurassic Park* and other bestsellers, has a background in medicine. He once wrote the following introduction for a paper published in the *New England Journal of Medicine* (1975):

> Most medical communications are difficult to read. To determine why, contributions to three issues of the *New England Journal of Medicine* were studied and the prose analyzed.

Crichton's first sentence awakens interest. It is succinctly written in only seven words. The entire section is short – a mere three lines long, and not a word needs to be added.

Here is another fine introduction, to a paper published in the *BMJ* (McGarry 1994):

> Nose bleeds in adults are the commonest reason for emergency admission to an otolaryngology ward, but the cause of the condition remains unknown.[1] Case reports suggest an association between nose bleeds and regular, high alcohol consumption.[2–5]
>
> We conducted a prospective case-control study to compare the alcohol habits of adults with nose bleeds with those of controls being treated for other otorhinolaryngological conditions.

These introductions, like many other well-written introductions, contain a brief description of two items:

(1)   the problem;

(2)   the proposed solution.

However, the first few sentences sometimes contain general, even vapid, statements, as in this example from a manuscript under preparation submitted by a course participant of mine:

> Respiratory diseases are important health problems throughout the world and often lead to morbidity and death.

These platitudes could be omitted ruthlessly, as could also the empty words of the next sentence. Only in the third sentence does the author come to the point:

> An important risk factor for developing Chronic Obstructive Pulmonary Disease (COPD) is chronic cigarette smoking (1).

I advised the author to open the introduction with the third sentence. The reference (1) should be to a carefully chosen review article describing the problem.

You probably need more than two or three sentences for your introduction, but it should preferably not exceed one page in length (typed double-spaced). More space may be required for certain topics, such as occupational science, medical ethics, and nursing and health care. Check the current version of the Instruction for Authors.

If you have previously published part of the work other than in the form of a congress abstract, you should say so in a few words at the end of the introduction.

How to Write and Illustrate a Scientific Paper

13

Methods

Methods are usually best described in the order in which they were used. So, in the following sentence, taken from a manuscript in preparation, it would be helpful if you reversed the order of presentation, as from:

> Cell growth was stopped with colchicine after incubation for 65–70 hours at 37 °C.

to:

> After the incubation of cells for 65–70 hours at 37 °C, their growth was stopped with colchicine.

Unless a previously published method is generally known, the reader will appreciate being told its essential features. Thus, a reference figure may well be considered inadequate, as in this example from a manuscript in preparation:

> Kidney volume was measured as previously described.[3]

Little need be added to give the reader the broad outline of the method (revised text in boldfaced italics):

> The kidney volume was measured *with an ultrasound apparatus containing a built-in volume program.*[3]

A new procedure, however, should be described in sufficient detail to allow a trained scientist to repeat the investigation. For the novice writer it can be hard to find a middle course between too much and too little information; often the novice errs on the side of too much. An experienced colleague can help to remove excess detail.

The subjects

I once saw a note taped to a door asking for volunteers – completely healthy, nonsmoking women aged 55 to 60 – to participate in a study of the effect of estrogen on blood circulation in the legs. The door led from a huge bicycle garage (reserved for workers at the hospital) into the elevator hall. In a subsequent report of this study each of these details has to be related, because bicycling women of this age might not be representative of the reader's patients – especially not concerning the main topic of the study, blood circulation in the legs.

In fact, most doctors reading your paper will ask themselves, "Does this apply to my patients?" To assess this, they need to know precisely the source of the participants and the details of the entry and exclusion criteria. If you are uncertain how to present it, I recommend as a model the section "Subject Recruitment and Enrollment Criteria" in a paper on treatment of the common cold in children, published in *JAMA* (Macknin *et al.* 1998).

Informed consent

Before ethics committees became common, lecturers teaching research methods, including the matter of obtaining informed consent, could get the following comment from course participants, "But if we ask the subjects they might say no" (Holmes 1997).

That's right. When patients have a free choice, few studies have a consent rate of 100 percent. So, if you are writing the often seen construction "All patients gave informed consent," give it a second look. Probably you are referring only to those patients who had already been enrolled in the study. If that is the case, say so.

If, on the other hand, you are referring to all those patients who fulfilled the criteria for being enrolled (called *eligible* patients), then the reader would be interested to know whether the patients really were informed of possible side effects of the trial, and if they were given a copy of the written informed consent form.

Normally, the following could suffice, although it is a minimum (Jha *et al.* 1998):

> Patients were informed of the purpose of the trial and had to give their signed informed consent before being enrolled.

Then state in the results section how many patients declined to participate. If you know their reasons for doing so, these should be reported too. But remember, patients are free to say "no" without giving a reason.

Let us now take a closer look at one particular type of study, the randomized controlled trial. When properly conducted, it is the most reliable way of comparing treatments. However, reports of such trials frequently omit important features of the study design. We shall consider some of them.

Omissions in reporting of randomized controlled trials

Randomization

Randomization means that subjects are allocated to the treatment and control groups by chance (at random). But it is inadequate just to say that a study was randomized, without telling how. The reason is that all methods of randomization are susceptible to conscious or unconscious manipulation by investigators, the degree of which varies from one method to the other. The two methods considered most reliable are the following: a telephone call to an independent center for a computerized "flip of a coin"; and the use of carefully prepared envelopes (sealed, opaque, identical, and serially numbered). Procedures easier to manipulate are alternate cases, odd and even birth dates, or file numbers. They should be avoided. So, to evaluate the trial properly, the reader needs to know how the assignment was made, as in this example (European Carotid Surgery Trialists' Collaboration Group 1998):

> We randomised . . . by telephone to the Clinical Service Unit in Oxford. A computer program generated the randomisation schedule . . . making it impossible for the local investigators to know whether the next allocation was going to be to surgery or control.

Blinding

Specify the *blinding* method used. For example, when three parties are involved – the patient, the treating physician, and the evaluator – some designate the study *triple-blind*, others *double-blind*. And what is implied in the term *double-blind* in the paper entitled "Double-blind study . . . on . . . rats," published in *Acta Chirurgica Belgica* in 1978?

Moreover, how should we denote blindness in a study in which only the evaluators are masked, as was the case in the famous trial of streptomycin in the treatment of pulmonary tuberculosis (Medical Research Council 1948). The term *blind* was not even mentioned in the paper, which merely described what was done:

> The [X-ray] films have been viewed by two radiologists and a clinician, each reading the films independently and not knowing if the films were of C [control] or S [streptomycin] cases. There was a fair agreement among the three; at a final session they met to review and discuss films on which there had been difference of interpretation, and agreement was reached without difficulty on all films.

This is a careful reporting of how blindness was accomplished.

How to Write and Illustrate a Scientific Paper

Number of subjects

Before you start the investigation, calculate the sample size needed to demonstrate a difference, if it exists. This calculation should be reported. Here is an example from a study on the healing of leg ulcers, published in the *BMJ* (Morell *et al.* 1998):

> *Sample size*
> To have an 80% chance of detecting as significant (at the 5% level) an increase in healing from 50% to 70%, 206 patients were required.

However, the number of patients needed in a study refers to the number of patients who complete the trial, rather than the number who start it. To add a calculated drop-out rate, say 10%, would therefore be advisable:

> To allow for a 10% drop-out rate, 230 patients were enrolled.

Sample-size calculation is so important a part of the study design that it deserves a separate subheading.

Let us turn briefly to animal studies. An example of a fine study design for a trial on dogs with perianal fistulas was reported by Mathews and Sukhiani (1997) in *Journal of the American Veterinary Medical Association*. The number of dogs needed was calculated, and the random allocation of the dogs to the treatment and control groups was described. In most studies on animals, such information is lacking.

A checklist

In 1994, two expert groups independently published detailed checklists for the reporting of randomized controlled trials. Two years later these groups produced a unified list, the CONSORT (Consolidated Standards of Reporting Trials) statement. It states in detail which items must be included in a report (Begg *et al.* 1996; Moher *et al.* 2001; CONSORT Website 2001).

14

Results

As the reporting of the randomized controlled trial covers most aspects of the *principles* of scientific writing, I have used it as a model for this chapter. This approach spares us tedious repetition. The chapter therefore has two parts: the flow of participants and their follow-up; and the outcome of the study.

Participant flow and follow-up

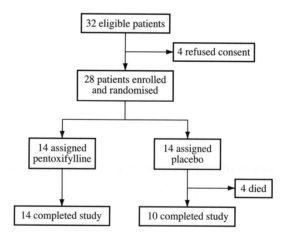

Figure 14.1 Trial profile. (Reproduced, with permission, from K. Sliwa, D. Skudicky, G. Candy, T. Wisenbaugh, P. Sareli, Randomised investigation of effects of pentoxifylline on left-ventricular performance in idiopathic dilated cardiomyopathy, *The Lancet* 1988; **351(9109)**:1091–3, © The Lancet Ltd.)

How to Write and Illustrate a Scientific Paper

This seemingly fine flow chart (14.1) would have been even more informative if it had included two additional boxes above the first one, giving the numbers of patients screened and excluded. Here is part of a flow chart with such information.

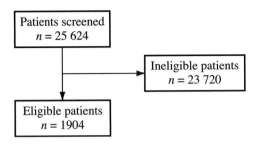

Figure 14.2 Part of a flow chart showing that no fewer than 93% of the patients screened were excluded. (Reproduced, with permission, from The Publications Committee for the Trial of ORG 10172, in Acute Stroke Treatment [TOAST] Investigators. *JAMA*, April 22/29, **279**:1265–72. Copyright 1998, American Medical Association.)

Only 1904 of the 25 624 potential subjects were selected. This information is useful to have in case you intend to make a confirmatory study. In the main text of this paper, the reasons for exclusion were given in detail. Such data can help readers to assess potential bias in patient selection.

Dropouts

A high rate of withdrawal (say 15 percent or more) can invalidate the conclusions of the study (Lang and Secic 1977, 24). The number of dropouts and their reasons for withdrawal should be reported for each group separately.

Moreover, dropouts should be included in the analysis of the study on an "intention-to-treat" basis. This is often neglected and can result in incorrect reporting. If you simply ignore the drop-outs and if, for example, the reason for their exclusion is chiefly the side effects of the new treatment tested, then the ensuing comparison would be biased in favor of this treatment. Hence, all randomized participants should be analyzed in the groups they were originally allocated to, including even those patients who did not receive the intended treatment and even those who, for some reason, subsequently received the treatment of the alternative group. The use of intention-to-treat analysis should be indicated in the flow chart, for example in either of the following ways.

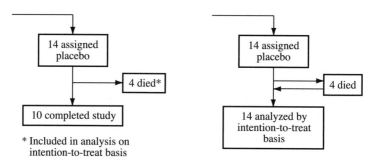

Figure 14.3 Alternative displays of the right arm of Figure 14.1 to indicate analysis on intention-to-treat basis

Loss of participants to follow-up

Participants who completed the treatment but who were lost to follow-up are likely to be atypical in critical ways. They could be patients who did not return because they had recovered, or because they had died, or who were still sick and did not wish to burden the doctor with an unsatisfactory outcome. Describe their characteristics as at the last examination.

Details of randomized participants

In a manuscript under preparation, I saw a detailed description of the study group, but the controls were presented as follows:

> A control group of sixteen healthy volunteers underwent investigation in the same manner.

How to Write and Illustrate a Scientific Paper

Such scanty presentation of the controls has been the reason for many a rejected paper. The controls should be described just as thoroughly as the subjects of the treated or exposed group, as in the following table from a study on the duration of pregnancy after laser conization of the cervix.

Table 14.1 Clinical characteristics

Characteristics	Cases (n = 64)	Controls (n = 64)	P
Age at delivery (y)	31.3 ± 4.19	31.4 ± 4.15	NS
Primigravida	35 (54.7%)	33 (51.6%)	NS
Previous preterm delivery	2 (3.1%)	3 (4.7%)	NS
Previous miscarriages	7 (10.9%)	8 (12.5%)	NS
Previous voluntary abortions	4 (6.3%)	2 (3.1%)	NS
Preeclampsia	2 (3.1%)	1 (1.6%)	NS
Operative vaginal delivery	13 (20.3%)	12 (18.8%)	NS
Cesarean delivery	9 (14.1%)	16 (25.0%)	NS

NS = not significant

Data presented as mean ± standard deviation or n (%).

Source: Reprinted from Raio *et al.*, with permission from the American College of Obstetricians and Gynecologists (*Obstetrics & Gynecology* 1997; **90**:978–82).

Be cautious, however, when comparing baseline characteristics with the use of statistical tests. A minor imbalance in a key prognostic factor can have a profound effect on a treatment comparison, even when the imbalance shows "no significance." Significance testing can thus obscure an important imbalance. Therefore, in addition to the statistical testing (some experts say "instead of statistical testing"), compare the baseline characteristics with the use of clinical experience – and common sense (Altman 1985; Bailar 1986). Then, state in the main text what you have found. The *P* values could thus have been omitted from this table and replaced in the main text with, for example:

> There were no *clinically* meaningful differences between the groups in their baseline characteristics.

Study outcome

Try not to repeat tediously in the text what is already clear from a perusal of the tables. The text should emphasize the important observations and present them in order of decreasing interest, beginning with the main finding. Tables are used for reporting the details of the outcome. The following text repeats the content of the table:

> As Table 1 shows, the mean ± SD of nocturnal plasma-melatonin concentrations was 19.0 pg/mL ± 11.9 in the 6 patients in the suicidal group and 45.5 pg/mL ± 27.1 in the 22 controls ($P < 0.05$).

There is no need for such repetition; merely state the main point:

> Patients with a history of attempted suicide had significantly lower nocturnal plasma-melatonin concentrations than did controls ($P < 0.05$) (Table 1).

An exact P value would have made the presentation even more informative. Below is part of an excellent table that presents both confidence intervals and exact P values, thus enabling readers to judge for themselves the clinical importance of the results.

Table 14.2 Outcome of treatment

Outcome	Number (%) of patients with outcome		Difference in percentage (95% CI)	P
	Ciprofloxacin n = 60	Pivmecillinam n = 60		
Clinical success	48 (80%)	39 (65%)	15.0 (− 0.7 to 30.8)	0.10
Bacteriological success	60 (100%)	54 (90%)	10.0 (2.4 to 17.6)	0.03
[etc.]				

Source: Reproduced, with permission, from M.A. Salam, U. Dhar, W.A. Khan, M.L. Bennish, Randomised comparison of ciprofloxacin suspension and pivmecillinam for childhood shigellosis. *The Lancet* 1998; **352(9127)**:522–7, © The Lancet Ltd.

How to Write and Illustrate a Scientific Paper

Be careful also in the reporting of side effects. It could be the most important part of the study. (Does the drug cause so much nausea that patients will not take it?) Therefore, describe adverse effects as thoroughly as beneficial ones. If no adverse reactions were found, say so.

15

Discussion

Toward the end of World War II, General Patton advanced with his tanks through the enemy lines. From motorcycle dispatch riders he received intelligence reports from other parts of the front. When a rider began reading out the dispatch from the very beginning, Patton most often asked him to go directly to the bottom line.

In the same way, hurried readers of scientific papers thumb through the pages to find the final paragraph of the discussion. Why? Apparently, because that is where the reader can usually get a comprehensive conclusion of the results.

But there is no generally accepted form on how to arrange the various parts leading up to the conclusion. So to help you I have chosen, as a model, an especially well-designed discussion in a paper by Logan *et al.* (1993). Based on this I have composed a structure for the discussion section that can be used as a guide when writing this section. It has four parts: *Main message, Critical assessment, Comparison with other studies*, and *Conclusions*. Let us discuss them step by step.

(1) **Main message**, which answers the question posed in the introduction section and includes the main supporting evidence.

How to Write and Illustrate a Scientific Paper

Most often, however, the opening paragraph of a discussion unnecessarily recapitulates in detail what the readers have already been told twice, once in the abstract and again in the results section, as in the following example. (Imagine General Patton listening to this!)

Discussion

Results of the first phase of this study show that men assigned to Hospital Corpsman and Mess Management Specialist occupations have the highest overall hospitalization rates across the three decades of a 30-year navy career. Rates also are elevated for the groups of Construction/Manufacturing, Deck, Ordnance, and Engineering/Hull while the lowest rates across the three decades are observed for the group of Miscellaneous/Technical, Electronics, and Administrative/Clerical.

Now, compare this with the next example in which Logan *et al.* open the discussion with an answer to the question posed in the introduction.

Discussion

Our data support the hypothesis that taking aspirin or other non-steroidal anti-inflammatory drugs protects against the development of colorectal cancer and suggest that it does so by reducing the prevalence of colorectal adenomas.

Having seen this, readers may well wonder how valid the paper's arguments are. So, here is the right place to present the strengths and weaknesses of the study.

(2) **Critical assessment**, that is, opinions on any shortcomings in *study design*, limitations in *methods*, flaws in *analysis*, or validity of *assumptions*.

Logan *et al.* gave primacy to this important part of the paper by giving it a separate heading:

> *Bias and confounding*
> Could bias or confounding account for these findings? . . .

Whether convinced by the arguments or still skeptical, readers will then want to know how the findings agree (or contrast) with previously published work.

(3) **Comparison with other studies**, where inconsistencies are discussed.

If you intend to discuss several observations, start with the most significant, continue with the next most important, and so on. Thus, beginning with your most significant finding, you start the comparison with studies whose results are largely consistent with your own. Then consider studies less compatible with yours, and so on. Conclude with any results that contradict your findings.

All the time, discuss similarities and differences. If you cannot explain conflicting evidence, you could suggest how the discrepancy might be resolved by conducting a new trial.

Having completed such a comparison, Logan *et al.* rounded off the discussion in a paragraph headed "Conclusions."

(4) **Conclusions,** that is, comments on possible biological or clinical implications and suggestions for further research.

> . . . Studies are now needed to confirm these findings, to determine how non-steroidal anti-inflammatory drugs might act, and particularly to see if [these] drugs can prevent the recurrence of adenoma or even cause sporadic adenomas to regress.

In the following I will set out further suggestions for writing the discussion section.

How to Write and Illustrate a Scientific Paper

Evaluate the results – not the authors

Consider this:

> A simple but very keen-sighted observation was made in the Gothenburg study, namely, that there was a relationship between the waist to hip circumference ratio and myocardial infarction. [My translation from the Swedish.]

In my view, the only thing these investigators did was to use a measuring-tape. So, in what way was this keen-sighted? Reading the measuring-tape? And why "simple but" instead of "simple and," as if "simple" were something negative? I would have deleted the first ten words: "The Gothenburg study showed that . . . "

Avoid claiming priority

In a discussion section, I once saw this:

> Our study appears to be the first one in which an open-end catheter method was applied to the study of tubal motility in the primate.

Some time later, in *Letters to the Editor*, a reader told us that several similar works on humans had been published and called the claim of priority "intellectual piracy." As the study was made on apes (*primates* include both humans and apes), the author of the paper replied:

> The quotation about being the first study in the primate is obviously wrong. We . . . should of course have said *nonhuman primate*.

Following this line of argument, most studies could be designated "the first," because most of them have a design of their own. Try instead to establish the novelty of your work by telling the reader in the introduction how the design of your study differs from that of previous works, as in this fictitious example:

> Most studies have been made on humans; ours was made on apes.

Then tell the reader why this approach ought to be superior.

The reference-13 trick

At this point, let us assume that you have completed your study after three years of hard work. When you now scan the latest issue of your target journal, you come across a paper on a subject very similar to yours; same questions, same answers.

You are horrified. Your first intuition is to ignore the article, "After all, editors and referees are too busy to have time to read journals." At that very moment, Reference-13 Himself approaches you and whispers in your ear: "Bury it in the discussion!" – meaning that if you refer to the paper in a short subordinate clause in the depths of the discussion section, it will hopefully pass unnoticed (*BMJ* editorial 1985). But the editor and referee will find it, for sure, because they know the trick.

Your citation should instead be inserted in the introduction section, as Reference 1 or 2. That doesn't mean that your cause is lost; your investigation could be an important confirmatory study.

Here is an example of a generous way of handling this problem. Two researchers, Karman and Potts (1972), had developed a surgical method. When they subsequently searched the literature, they found that a similar technique had been described 35 years earlier in a Ukrainian journal in the Russian language with no English summary. They honestly gave full credit to the initial inventor:

> Since the development of this apparatus it has come to our attention that Bykov developed an analogous procedure in 1927.[2]

16

Acknowledgments

You should not forget to thank the people who have really helped you, but whose contributions do not justify authorship. But be specific. When I saw this acknowledgment, I wondered what exactly these persons had done (family names are fictitious):

> We thank C. Roe, D. Doe, and S. Poe.

In the following example it is evident what each person did:

> We thank Betsy Roe and Gerri Doe for their assistance in preparing the data; William Poe for the medical photography; Marian Loe and David Coe for their critique of the findings of this study; and Fred Noe for reviewing the 200-μm and 400-μm specimens.

But take care not to give others credit for your own work. Otherwise the reader will wonder what *your* contribution was.

The often used "wish to thank" can be shortened to "thank." Avoid using professional or courtesy titles in the acknowledgment. The example above was written accordingly. It also tells us the given name of those acknowledged.

Persons you want to thank should be asked if they are willing to be acknowledged and if they approve the wording you have used to mention them. This is because colleagues who have read and corrected your manuscript may disagree with some of its central points. To acknowledge them could imply their approval of the content of the paper.

Always give credit for financial support

When you thank sources of financial assistance, be careful how you present their names. If, as in this case, the name of the funding agency is not in the language of the journal:

> The study was supported by "Kronprinsessan Margaretas Arbets-nämnd för synskadade."

you should use either a translation alone:

> The study was supported by Crown Princess Margareta's Working Group for the Visually Handicapped.

or, as a courtesy to the granting authority, both the original name and, in brackets, the translated name:

> The study was supported by "Kronprinsessan Margaretas Arbets-nämnd för synskadade" [Crown Princess Margareta's Working Group for the Visually Handicapped].

But do not thank all grant-giving agencies who have supported your research work over several years past. Thank only those which supported the study you are now reporting because, just as research teams compete with one another, so too do granting authorities. And if one authority alone has chosen and supported a promising work, you should give that authority all due credit.

Note that some journals ask that funding organizations be named on the title page instead of being included in the acknowledgments.

17

References

At one time, there were over 250 different styles of reference in the scientific literature (Garfield 1986). The editors of some major biomedical journals therefore had good reason to convene in Vancouver, Canada, in January 1978, to work out a uniform reference style. One of their suggestions was that authors should number references in the order in which they appear in the text (International Committee of Medical Journal Editors 1997).

Vancouver versus Harvard style?

Although many of the major journals in the biomedical field have adopted the Vancouver style, some still prefer the Harvard system (first used in 1881 by a zoologist at Harvard University [Chernin 1988]) in which the author's name and the year of publication are cited in the text. In the fictive sentence below, I have mixed the two styles to illustrate their differences:

> A reference figure (17) in the Vancouver style says less than a name-and-year reference (Einstein 1941) according to the Harvard system.

Most readers prefer the Harvard system because they like to know just what author is being cited as they read the text. Still, the name-and-year system does have disadvantages: difficulty for readers who see an interesting item in the reference list in locating that reference in the main text; and, more important, the disruption of the text when a large number of references need to be cited within a paragraph, as in this example (Bengtsson 1968):

> This method was introduced by Aburel in 1938, but he was followed by only a few workers in the succeeding 20 years (Bommelaer 1948; Cioc 1948; Kosowski 1949; de Watteville and d'Enst 1950). During the 1960's however, hypertonic saline has been increasingly employed (Bengtsson and Csapo 1962; Jaffin *et al.* 1962; Wagner *et al.* 1962; Larsson-Cohn 1964; Møller *et al.* 1964; Sciarra *et al.* 1964; Wiqvist and Eriksson 1964; Bora 1965; Short *et al.* 1965; Turnbull and Andersson 1965; Wagatsuma 1965; Cameron and Dayan 1966; Gochberg and Reid 1966; Klopper *et al.* 1966; Christie *et al.* 1966; Ruttner 1966; Olsen *et al.* 1967).

Using the Vancouver system, the text above can be condensed to about one-third of its original length:

> This method was introduced by Aburel in 1938,[1] but he was followed by only a few workers in the succeeding 20 years.[2–5] During the 1960s, however, hypertonic saline has been increasingly employed.[6–22]

As shown, an important name and year can be featured within the text even when using the numbering system. Nevertheless, you have to follow the style of the journal to which the paper is to be submitted. So read the current version of its Instructions for Authors.

Now at last you are ready to submit your paper. At this moment you notice that you have used an inappropriate citation system. Today, this is not a catastrophe. By using a suitable computer program you can, at the touch of a few keys, produce reference lists in the format of your choice and, in the text, substitute names for numbers (or vice versa).

In two pages, I would like to show in detail how references are written in the main text and in the reference list according to the Vancouver and the Harvard systems. I also take the opportunity to show how to refer to unpublished results and personal communications.

The Vancouver system – also called the numbering system

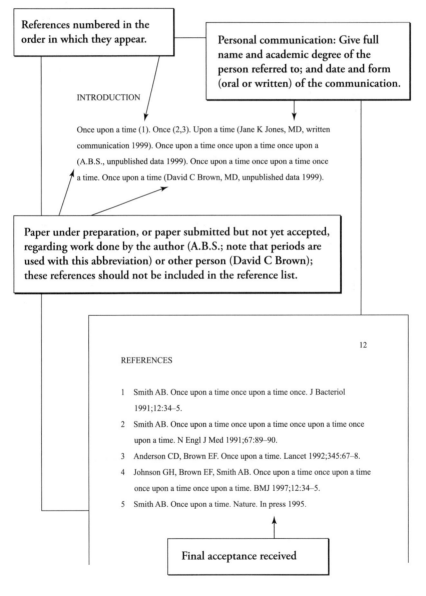

The Harvard system – also called the name-and-year system

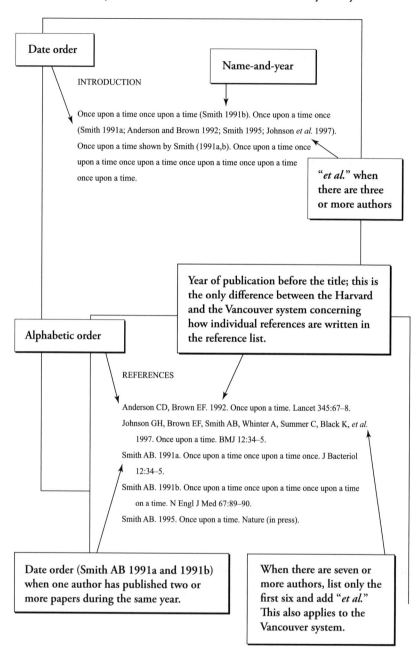

Date order

INTRODUCTION

Once upon a time once upon a time (Smith 1991b). Once upon a time once
(Smith 1991a; Anderson and Brown 1992; Smith 1995; Johnson *et al.* 1997).
Once upon a time shown by Smith (1991a,b). Once upon a time once
upon a time once upon a time once upon a time once upon a time
once upon a time.

Name-and-year

**"*et al.*" when
there are three
or more authors**

**Year of publication before the title; this is
the only difference between the Harvard
and the Vancouver system concerning
how individual references are written in
the reference list.**

Alphabetic order

REFERENCES

Anderson CD, Brown EF. 1992. Once upon a time. Lancet 345:67–8.

Johnson GH, Brown EF, Smith AB, Whinter A, Summer C, Black K, *et al.*
1997. Once upon a time. BMJ 12:34–5.

Smith AB. 1991a. Once upon a time once upon a time once. J Bacteriol
12:34–5.

Smith AB. 1991b. Once upon a time once upon a time once upon a time
on a time. N Engl J Med 67:89–90.

Smith AB. 1995. Once upon a time. Nature (in press).

**Date order (Smith AB 1991a and 1991b)
when one author has published two or
more papers during the same year.**

**When there are seven or
more authors, list only the
first six and add "*et al.*"
This also applies to the
Vancouver system.**

How to Write and Illustrate a Scientific Paper

Accuracy of references and quotations

When I was a novice editor, I endeavored to evaluate personally the content of each submitted paper (in addition to the assessments made subsequently by referees). If I didn't know the subject well enough, I marked the key citations in the reference list when reading the text and then tried to retrieve them.

I was surprised to find how often I was unable to track down an article. In some submitted papers as many as every second reference was untraceable by direct search. Through the journal's volume index, however, I found some of them, but under the wrong year, wrong volume, or wrong page. Other errors found in retrieved publications included incorrect title, wrong author(s), misspelling of author's name or, more seriously, misquoting of other authors' results.

What can an author do to make citations more accurate? Most misquotations can be avoided by rereading the publications cited. Thus, never rely on memory alone! The number of errors can be reduced by checking and rechecking not only new references but also those lifted from your own computer bank of citations. Even references downloaded from MEDLINE can conceal errors! This means that you should always have a copy of the publications to be cited – at hand. (Some journals require that authors send, with the submitted paper, a photocopy of the first page of every reference listed!)

Quoting from another article is allowed only if the original publication is unavailable. However, citing a publication you have not seen can be hazardous, as the following illustrates. For nearly 50 years, several authors of English-language papers referred to a Dr. O. Uplavici (Dobbel 1938). He was assumed to be the author of a Czech-language article, reporting the first experiments in which amebic dysentery was transmitted from man to cat. Actually, "O úplavici" was the title of the paper and means "On dysentery." The first author who cited the paper, Kartulis (1887), made the

(Tietung Hospital 1975)

In Vancouver style:

In 1975, Tietung Hospital (7) reported…

In your reference list it would appear as (Harvard style):

Tietung Hospital. 1975. Fetal sex prediction by sex chromatin of chorionic villi cells during early pregnancy. *Chin Med J* **1**:117–26.

An unsigned editorial can be referred to in the text as in the following example:

(*Nature* editorial 2006)

and in the list of references:

Nature editorial. 2006. Save the lungfish. An Australian dam project threatens a living fossil. **442**:224.

Editorials or other articles by unknown authors can also be referred to by citing in the text the first few words of the title:

(Save the lungfish 2006)

and in the list of references:

Save the lungfish. An Australian dam project threatens a living fossil [editorial]. 2006. *Nature* **442**:224.

In general, citing an author as *Anonymous* is to be avoided in scientific writing.

Record titles in the language of publication

I once saw a manuscript in which the first seven references listed were written in five different languages: Bulgarian, Hungarian, Italian, German, and English. Readers are best served if only one language is used, that is, the language of the journal in question, which its readers can be assumed to understand. Let us look at one of the references mentioned, that in Bulgarian:

> Tanchev S, Asparuhov A, Tanchev P, Gramcheva O. Vurkhu vuzstanoviavaneto sled rodova fractura na kliuchitsata. Akush Ginecol (Sofia) 1987; **26**:49–XX.

As the language of the journal is English, the Bulgarian title needs to be translated into English. The translated title is then placed in brackets and additional information in parentheses. (Like the original reference, the following one is written in Vancouver style.)

> Tančev S, Asparuhov A, Tančev P, Gramčeva O. [Healing of fractured clavicle in newborns] (In Bulgarian with English abstract) Akush Ginecol (Sofia) 1987;**26**:49–XX.

Note that I have placed a diacritical mark over the letter *c* (replacing *ch*) when rewriting the names of the authors, which is the way they themselves have written their names in the English abstract. To do so is a courtesy. If your computer cannot cope with this mark, insert it by hand, indicate the change in the margin, and inform the editor about it in the covering letter.

You can also provide both a non-English-language title and the translation:

> Svedin G. Transkutan nervstimulering som smärtlindring vid förlossning. [Transcutaneous electrical nerve stimulation for analgesia in childbirth.] (In Swedish with English abstract.) Läkartidningen 1979;**76**:1946–8.

Names of journals

Abbreviate journals according to the listing in *Index Medicus* (www.nlm.nih.gov). If a journal is not listed (about three in four biomedical journals are not), spell out the journal's title in full.

How to refer to World Wide Web

Because a website may be updated after you have seen it, even disappear, you should give the date you accessed the site and also keep a printed copy of it. No firm rules exist yet for citing electronic materials. Consult the Instructions to Authors of your target journal. In this book I have used the following format in most cases:

> Animal Info. 2002. Information on rare, threatened and endangered mammals. Severna Park (MD): Animal info. www.animalinfo.org (accessed 2 February 2002).

18

Ph.D. and
other doctoral theses

The Ph.D. (Philosophiae Doctor) is the highest university degree. It is acquired after writing a doctoral thesis (or dissertation) and defending it at an oral examination.

There are almost no generally accepted rules for thesis preparation. The structure varies from country to country, "from institution to institution and even from professor to professor in the same department of the same institution" (Day and Gastel 2006).

However, theses are of two kinds: monographs and compilations of articles. Monographs are the most common form, especially in the humanities, theology, and law. But compilations are increasingly used in medicine, technology, and the natural sciences.

Compilations: the theses of the future

Compilations are based on articles that have "been scrutinized by international peer review, probably more prestigious than local committees" (Carling 2006). They are recognized in Argentina, Australia, India, Japan, the United States, and in such European countries as the Czech Republic, Finland, Germany, the Netherlands, Norway, Spain, and Sweden; they are permitted in the United Kingdom, but are not common there (Burrough-Boenisch 2006).

Compilations are of two types. In one, the reprinted articles are sandwiched between introductory and concluding chapters. In the other, the reprinted articles are appended to a summary of their contents. In both types, the articles are published or publishable in refereed journals and often have several authors, with the doctoral student as first author of most of them. The subject of this chapter is the type of compilation that includes a summary; for lack of an official name, I propose to call that summary a thesis overview.

Many of the details described in other parts of this book apply to the writing of a thesis in general. So, to avoid tedious repetition, I will discuss only new features here. The advice is applicable also to other doctoral and postgraduate degrees.

Contributors

Most research is nowadays performed in groups. You must do your utmost to indicate clearly what parts of the work were yours. The examiner will be curious to know how much you contributed to the study design, data collection, data analysis, and, especially, writing of the manuscript. Table 18.1 is an exemplary presentation of a list of contributors to the papers of a thesis, shown as a facsimile (Theander 2005). See the next page.

Table 18.1 Detailed list of contributors to the papers of a doctoral thesis (facsimile from Elke Theander 2005, with permission)

Living and Dying with Primary Sjögren's Syndrome		11

Contributions to the papers

Study design	Paper I	Elke Theander
		Ingrid Nilsson
	Paper II	Rolf Manthorpe
		David Horrobin
	Paper III	Elke Theander
		Lennart Jacobsson
	Paper IV	Elke Theander
		Lennart Jacobsson
	Paper V	Elke Theander
		Lennart Jacobsson
Data collection	Paper I	Ingrid Nilsson
		Elke Theander
		Rolf Manthorpe
	Paper II	Elke Theander
		Rolf Manthorpe
	Paper III	Elke Theander
	Paper IV	Elke Theander
		Rolf Manthorpe
		Otto Ljungberg
	Paper V	Elke Theander
		Rolf Manthorpe
Data analysis	Paper I	Elke Theander
	Paper II	Elke Theander
		Jan Åke Nilsson
	Paper III	Elke Theander
	Paper IV	Elke Theander
		Anna Bladström
	Paper V	Elke Theander
		Henrik Månsson
	All papers	statistical advice by Jan Åke Nilsson
Manuscript writing	All papers	Elke Theander
Manuscript revision	Paper I	Lennart Jacobsson
		Rolf Manthorpe
		Ingrid Nilsson
		Torkel Wadström
	Paper II	Rolf Manthorpe
		David Horrobin
		Lennart Jacobsson
	Paper III	Lennart Jacobsson
		Sven Ingmar Andersson
		Rolf Manthorpe
	Paper IV	Lennart Jacobsson
		Gunnel Henriksson
		Otto Ljungberg
		Rolf Manthorpe
	Paper V	Lennart Jacobsson
		Rolf Manthorpe
Language revision	Papers	Dorothy Björklund
	Thesis	Helen Sheppard

Thesis at a glance

Thesis at a glance could be described as abstracts of the abstracts and is extremely helpful to the reader. Here is an example of one part of it (Theander 2005):

> **Paper II – Does treatment with gammalinolenic acid (GLA) alleviate fatigue and glandular dysfunction in primary Sjögren's syndrome?**
>
> **Patients:** 90 patients (+30 in pilot trial).
>
> **Methods:** Double-blind placebo-controlled randomized trial.
>
> **Conclusions:** GLA [had] no effect on fatigue or glandular signs and symptoms.

General introduction

You may well have come to know more than anyone else on the planet about your narrow subject, so your general introduction could and should be a highly readable piece of work. You may, if relevant, go back millions of years in the opening phrases, as in this introduction to a thesis on the use of ultrasound in medical diagnostics (Andolf 1989):

> For millions of years, bats and dolphins have used ultrasound as a method for localization. It was not until 1912, when the Titanic catastrophe occurred, that scientists proposed that man as well should use ultrasound ... In Lund in 1953, the cardiologist Inge Edler [and] the physicist Helmut Herz [made observations that led to] the application of ultrasound in the medical field.

The thesis overview should be intelligible even to a non-specialist. Baranto (2005), for example, devoted 18 pages to introducing the reader to the anatomy of the spine, its function and possible degenerative changes. This was illustrated by drawings borrowed from a textbook. Such an approach is highly recommended.

Aims

One short sentence for each aim is often sufficient. Do not use abbreviations here; if you do, explain them. A non-specialist reader would probably find this aim (quoted from a thesis) unclear:

- to elucidate the [presence of] urogenital carriage of GBS in mothers of GBS infected infants and/or in GBS colonized women giving birth to neonatally healthy infants, with respect to . . .

. . . but not this (Christensen 1980):

- to elucidate the presence of group B streptococci (GBS) in the urogenital tract among adults;

Methods and results

I had just completed the papers for my thesis, in May 1973. Now, it was time for the overview. A senior colleague of mine suggested that I should take the papers, a pair of scissors, and a roll of tape in one hand and a bottle of brandy in the other, find a quiet place and cut-and-paste the overview. (We had no word processors in those days!) His advice had a point; the methods and the results sections could have been presented in that way. But if you take parts from your papers you have to paraphrase them, that is, present them in a new way. If you do not, you need quotation marks around the borrowed lines. Paraphrasing is difficult, especially concerning methodology, so I suggest the following instead.

As you have already described in detail the methods and the results in your papers, the specialist readers will, in the printed thesis, have access to them there. Instead, in the overview, present these sections on a non-specialist level. Choose words similar to those you would use in discussing these sections over lunch with a colleague from a speciality other than your own.

If you have one common group of subjects or samples, use a flow chart to show how they were allocated to the different studies. If you have two common bases, create two flow charts. Circles can be used in a similar manner; the size of the circles corresponds to the number of subjects or samples. Overlapping circles can be used to indicate that some studies had a part in common. Use a table to present studies based on different subjects or samples. Design simple drawings to explain your methods. For the results, it may be useful to construct tables synthesizing data from more than one paper.

Color photographs or diagrams are said to enhance an article, but they must add something, not only be decorations. If you have color photos or figures in your journal papers, the less expensive black-and-white photos will suffice in the thesis overview – if you find it necessary to reproduce them there at all.

General discussion

The general discussion in a thesis requires a slightly different approach from the discussion section of papers, in that you must consider both the whole picture and the individual pieces. Use a new subheading for each individual piece. Open the general discussion by explaining how you achieved your aims. In one thesis (Bergström 1994), one of the aims was:

> • To develop a technique for subretinal transplantation of retinal cells . . .

and the general discussion began:

> ### The transplantation procedure
> The subretinal transplantation technique that we have developed has turned out to be easy to use and [gave] good and consistent results.

What makes the writing of this sentence especially good is that the author used almost exactly the same wording as in the sentence describing the aim.

Acknowledgments

If you suffer from writer's block at the end of your writing, these acknowledgments could be a way out (family names fictitious; Azem 2005):

> **Acknowledgments**
>
> **Thanks** . . .
>
> My supervisors Ann-Mari Soe and Samuel Loe for supporting and inspiring me, and sharing their great knowledge in mucosal immunology. Present and former colleagues, collaborators, students, professors, and administrators & technical staff for [contributions to] this thesis.
>
> This study was supported by grants from . . .

This is better than skipping the acknowledgments; only once have I encountered a thesis without an author's thanks.

Most acknowledgments, however, are written in the traditional manner where you thank people individually. But, instead of merely thanking a coworker "for helping me with a lot of work," tell the reader precisely what she did, "Ewa . . . for showing me how to perfuse rat livers." Though most authors create their own illustrations these days using computer software, a few turn to a professional artist. If so, do not forget to thank him or her. I remember an artist showing me a textbook in botany that she had illustrated. She put her soul into it, she said; but she was not mentioned in the acknowledgments. A statistician would also appreciate a word, "Jan Doe, for guiding me through the jungle of statistics," as would the language corrector.

You are your own editor of your thesis. You may acknowledge people without their permission, contrary to the custom in papers.

But you must be absolutely certain that the person acknowledged would have permitted a particular wording, such as:

. . . Yvette Soe, for preparing good fillets of beef; . . .

Cover illustration

If you have a cover illustration, explain it and credit the photographer or the artist, on one of the first pages. Here is an example (Naylor 2005):

Cover picture: A [. . .] rat that has been under too much stress (left). In comparison, an [. . .] exercising rat that is alert and ready for the next challenge.
Illustration by Joen Wetterholm.

. . . and here is the illustration. The two rats illustrate the essence of the thesis.

Figure 18.1 Cover illustration. (Reproduced from Andrew S. Naylor 2005, with permission from the author; illustration by Joen Wetterholm, JoenArt.)

How to Write and Illustrate a Scientific Paper

A fresh look

I took a fresh look at 100 theses in biomedicine published in Sweden in 2005 and found the following trends. The average thesis was based on four papers or articles. The doctoral student was usually listed first among the authors; the main supervisor, last. Few papers had more than six authors. Of the four papers or articles, two were published and two were *in press*, submitted or *in manuscript*. Roughly, the thesis overview was 55 pages long (reference list excluded) and cited on average 170 bibliographic sources.

The structure of the thesis overview

Most types of scientific writing are highly structured. Thesis writing is not. So you are free to begin with the parts most likely to be read first: *Abstract, Summary in your native language, and Acknowledgments.* Give each of them a whole sheet of paper and, if you can restrain yourself, use only one side of the sheet, leaving the back blank. It will give your dissertation an inviting opening. This approach I found in a technology thesis (Synnergren 2005). In the following, you will find a list of headings for the thesis overview and how to order them. After the presentation of the list, I will give some further comments on it.

Abstract
Summary in your native language
Acknowledgments

Contents
Abbreviations
Glossary (or Definitions)
List of papers
Contributors to the papers
Thesis at a glance

General introduction
Aims
Methods (or Materials and methods)
Results (or Results and comments)
General discussion
Conclusions
Clinical implications [if relevant]
Implications for further research
References
Papers I–

The heading *Glossary* denotes here a list of words that are not explained in the thesis, because they are so well known to specialists such as one for example, *Glial cells*. A reader outside the speciality would appreciate being told that it means, *Cells in the supporting structure of nervous tissue.* This example was taken from a two-page Glossary of a thesis (Andersson Grönlund 2005).

In *Definitions* you can explain, for example, *Preterm delivery*, the meaning of which is evident to any obstetrician but not to every non-specialist; thus denoting, *Delivery before 36 completed weeks.*

I have already described and highly recommended *Contributors to the papers* and *Thesis at a glance.*

In the heading *Materials and methods*, *Materials* should be replaced by *Patients* or *Subjects* when appropriate as human beings are not referred to as material.

19

Letters and case reports

The thalidomide letter

The X-ray image of a woman in late pregnancy showed a fetus without arms. "Once in a lifetime, we are supposed to see something like this, but I have seen it twice in a couple of months," said one radiologist. Rumors of an increase in similar defects were frequent – the cause unknown.

About two months later, in December 1961, a letter appeared in *The Lancet* stating that women receiving thalidomide in early pregnancy for morning sickness frequently had babies with missing or deformed limbs (McBride 1961). The letter, comprising only 15 lines, ended with the following question:

> Have any of your readers seen similar abnormalities in babies delivered of women who have taken this drug during pregnancy?

The response was overwhelming. Eventually more than 10 000 babies in almost 50 countries were born with such defects (Thalidomide UK 2006). After the publication of the letter, the drug was immediately withdrawn worldwide. This letter counts as the first milestone for *The Lancet* (www.thelancet.com, accessed 9 December 2006) since its announcement of the value of penicillin in 1940.

Format and size of a letter

Just as in the thalidomide letter, you should be brief and to the point. The format of a letter is almost always the same: a title; a salutation, for example, "To the Editor"; the letter itself without subheadings; and a list of references. You may include one figure or one table, though such additions will be published only if they add substantially to the letter. The length of a letter, however, varies widely from one journal to another. Read the Instructions to Authors for further information.

Letters are of two kinds: those discussing recent articles and those describing preliminary research. A letter discussing a recent article is usually sent by the editor to the authors of the article discussed, and both the letter and any response will be published together. Research letters are brief reports of novel findings that might stimulate further research.

Transforming a paper into a letter

It was a clear case of hubris. I had submitted a report in the form of a full paper on a single patient to *The Lancet*. Of course, the paper was rejected. But the editor offered me and my co-workers the opportunity to have it cut to a letter.

The original paper had 12 pages of running text, two tables, and 14 references. We decided to accept the letter format only on condition that the journal managed to include every important piece of information. *The Lancet* did (Hoyer *et al.* 1979). The huge reduction of text from an original paper to a letter tells us that the format of original papers may sometimes be an uneconomical form of presentation.

How to Write and Illustrate a Scientific Paper

Case reports

Ideally, a case report should suggest a hypothesis that can be tested by others. For example, the associations between estrogens and endometrial carcinoma (Fremont-Smith *et al.* 1946) and between birth control pills and high blood pressure (Woods 1967) were first suggested by isolated case reports that led to controlled research and confirmation of the initial hypotheses.

Unfortunately, most case reports submitted to journals are just another observation of an unusual condition already well known. The case should instead have been presented at a departmental seminar.

The common subtitle "A review of the literature" is inappropriate, because a case report is too brief (usually only two pages of running text) to permit even a mini-review. Couldn't the young author (it often is a junior) write a review article separately? No, an acknowledged expert should write the review, often after an invitation. He or she is supposed to have the experience to evaluate the articles, emphasize the good ones, merely mention others, and, above all, have the courage to exclude works that are below standard. Reviews may include hundreds of references and are often used when you have to limit your bibliography – and you must be able to rely on them.

So which case reports do reach the stage of peer review? Most likely those that formulate a testable hypothesis, or those that have something to add – for example, a new diagnostic tool or a new treatment. And, of course, a case report is appropriate if a new phenomenon has been observed.

Format and size of case reports

If you have such a case to report, do it. The structure and size vary widely from one journal to the other. Consult the Instructions to Authors of the journal you choose. Usually, after a short introduction, the case (headed: "Case report") is presented, followed by a short discussion and a list of references. Roughly, the length is limited to two pages (double spaced) of running text, five references, and one figure or one table. To save space, present only relevant findings and do not give whole strings of normal serum electrolytes and white-cell counts. To limit the extent of the bibliography, cite references to comprehensive reviews of the literature.

A case report was rejected on formal grounds as it had six figures whereas the journal allowed only one. The author was quick to resubmit the manuscript with the same figures, although he had changed Figures 1–6 to Figure 1, a–f! This was not a joke; the author probably thought he could cheat an overworked editor with this trick.

Another case report dealt with an observation made during a routine abdominal operation. The report shed new light on such cases as would merit publication. However, the report had seven authors – too many to find room at the operating table. The manuscript was returned to the corresponding author who was asked to declare the contribution of each author. Instead, he resubmitted the manuscript with five of the seven authors omitted. The manuscript was accepted. A case report seldom needs more than two authors, one who made the observation and, if necessary, one who monitored the writing.

How to Write and Illustrate a Scientific Paper

20

Numbers

It has long been the custom to spell out numbers below 10, as shown in this example from *Newsweek* (Ridley 2003):

> . . . a rat has seven neck and 13 thoracic vertebrae, a chicken 14 and seven . . .

But authorities on *scientific* style now agree that all numbers should be expressen in numerals, rather than in words, in most circumstances:

> . . . a rat has 7 neck and 13 thoracic vertebrae, a chicken 14 and 7 . . .

However, we should still spell out numbers that begin a sentence. The following example is from the abstract of a published paper:

> Three thousand eight hundred and seventy-six mothers were examined by ultrasound at 7–12 weeks of gestation. One hundred and sixty-six (4.3%) were found to have a dead fetus.

But many readers find it difficult to grasp large numbers written in words, as in the example shown. Note how much easier to comprehend the passage becomes when it is recast so that the numbers fall somewhere in the middle:

Ultrasound examination of 3876 women at 7–12 weeks of gestation showed that 166 (4.3%) had a dead fetus.

Two numbers side by side

Placing unrelated numbers next to each other confuses the reader, as in this example taken from Mosteller (1992):

This group of patients with leukemia had an average white-cell count of 257, 112 lymphocytes and 145 other types.

Separate the numbers:

This group of patients with leukemia had an average white-cell count of 257, of which 112 were lymphocytes and 145 other types.

Here is another confusing construction:

2 500-mg tablets

Spell out the number easier to express in words and leave the other in numerical form:

two 500-mg tablets

Decimal point

The decimal sign in English is a point, not a comma:

0.3   (*not* 0,3)

Use a zero before the decimal point:

0.3   (*not* .3)

Thousands

American and British practice has been to indicate thousands with commas. In many non-English speaking countries, however, the comma serves as a decimal marker. To avoid confusion for an international readership, the Council of Biology Editors' Style Manual Committee (1994, 196) recommends the use of a space to mark off thousands in English writing:

12 345   (*not* 12,345)

Not all journals have adopted the practice of spacing. In deference to the editors of such a journals, follow the house style.

Numbers with several zeros

Modern *standard units of measure* go up and down in steps of 1000. An appropriate unit to remove surplus zeros is therefore easy to find:

3 µL   (*not* 0.003 mL)

In other cases, multipliers (exponents) can be used:

1.6 x 10^9 bacteria per mL

although multipliers should be avoided if they can be easily replaced:

12 million inhabitants   (*not* 12 x 10^6 inhabitants)

but never use "billion" in a scientific paper; it means 10^9 in the USA, but 10^{12} in most European countries.

Quotients of units

You are allowed to use *one* slash (/) to express quotients of units:

km/h

but not two or more. Thus, "milligram per kilogram per hour" is preferably presented by means of negative exponents:

mg·kg⁻¹·h⁻¹   (*not* mg/kg/h)

The use of negative exponents may be unfamiliar to some readers. For example, a veterinarian seeing that cows were fed 10 kg·day⁻¹ suggested that they were probably fed at night (Lindsay 1989).

However, the use of negative exponents has come to stay and we have to get accustomed to it.

Percentages

In a manuscript in preparation I once read that five percent of the patients with claudicatio intermittens had home-help service once a week. How, I wondered, could the community afford this luxury for people with a relatively moderate handicap. Then I found that the total number of patients studied had been no more than 20. So, in fact, only one single patient had received the service!

That mode of presentation contained two errors: the original data were missing, and the number of patients was too small to warrant expression as a percentage.

Here are some conventions of scientific writing concerning the use of percentages:

(1) If the total number is less than, say, 25, percentages should not be used at all.

(2) If the total number is between 25 and 100, percentages should be expressed without decimals (7 percent, not 7.2 percent).

(3) If the total number is between 100 and 100 000, one decimal may be added – and only one (7.2 percent, not 7.23 percent).

(4) Only if the total number exceeds 100 000 may two decimals be added (7.23 percent).

(5) The original data should always be included:

> Death occurred in 209 (7.2%) of the 2901 patients.

Note that the percentage is reported in parentheses, to give primacy to the original data. Thus, not 7.2 percent (209).

(6) The original data should never be presented with a slash construction. Thus, not 209/2901 (7.2 percent).

Rounding to two significant digits

Ehrenberg (1977) maintains that numbers are easier to compare after rounding to two significant digits. (Final zeros do not matter, as the eye can readily filter them out.) Compare these two statements:

(1) Between 1970 and 1975 the number of legally performed abortions in Sweden increased, from 17 134 to 33 926.

(2) Between 1970 and 1975 the number of legally performed abortions in Sweden increased, from about 17 000 to about 34 000.

In the second statement, the numbers have been rounded to two digits, and the two-to-one relationship between them is much clearer than in the first statement. However, when exact values for numerical data matter, such a drastic rounding off is not recommended, but can be used, say, in the discussion. (See also Chapter 8, "Rounding off.")

Enumeration

Numerals within parentheses are used to enumerate items, as in this example (*Animal Info* 2002):

> *Animal Info* (2002) lists three mammals as endangered in Afghanistan: (1) snow leopard, (2) markhor, a member of the goat family (not recently confirmed), and (3) tiger (may be extinct here).

However, if references in the text are numbered, italic letters must be used instead, to avoid confusion:

> *Animal Info* (1) lists three mammals as endangered in Afghanistan: (*a*) snow leopard, (*b*) markhor, a member of the goat family (not recently confirmed), and (*c*) tiger (may be extinct here).

What do we mean by "often"?

When presenting numerical data in text, readers feel more at ease with prose description than with actual numbers. But be careful! When 51 researchers fluent in English were asked to quantify the term *often*, they suggested a rate somewhere between 28 and 92 percent (average 59 percent; Toogood 1980)! So, nonnumerical expressions alone should best be avoided. Thus:

> Most of the patients (82%) . . .

21

Abbreviations

Abbreviations should be kept to a minimum. So a formulation such as the following is *not* to be recommended (quoted from Spiers 1984):

> ... a patient with ASHD and PHMI, SPCABG, who PTA for ERCP had an episode of BRBPR.

It requires some years in the profession to grasp immediately that this patient with atherosclerotic heart disease and a history of myocardial infarction, status post-coronary-artery-bypass graft, had an episode of bright red blood per rectum prior to admission for endoscopic retrograde choledochopancreatography!

The abbreviations used in this sentence are probably all accepted in the specialty. But just because an abbreviation is permitted does not mean that you are obliged to use it.

So when should you consider using an abbreviation? Let us take an example. The term *nonsteroidal anti-inflammatory drug* (accepted abbreviation NSAID) may not warrant abbreviation unless it occurs, say, a dozen times in a paper of standard length.

Some abbreviations are more readily understood than the full forms: DNA, AIDS, laser. Often such abbreviations are accepted in the major bibliographic databases. If so, you are free to use them without definition – even in the title and the abstract section.

Refrain as far as possible from inventing your own abbreviations. Try instead to find substitute expressions. Assume that you have made a study of young mature Sprague Dawley rats. You are now writing the paper and need to refer frequently to this group. You therefore consider devising a more convenient construction, such as the YMSD rats. Forget it! The editor will never accept it. So, what to do? If there are no other rats mentioned in the paper, just simply use "the rats"; otherwise, for example, "experimental" or "treated" rats.

Units of measure

Units are abbreviated when they follow a numeral. Otherwise, they are spelled out:

> 2 mg (but *two* milligrams)

Singular and plural have the same abbreviation:

> 1 wk   6 wk (*not* 6 wks)

An abbreviated unit takes no period unless it ends a sentence:

> mo (*not* mo.)

Abbreviated units need no explanation.

The title of the film *48 HRS.* (1982) – the movie with Nick Nolte and Eddie Murphy in the leading roles – is fine. But in scientific writing you should use *48 h*. Note that *hours* is abbreviated with a lower case letter. The capital H is the symbol for *hydrogen*.

By the way, *Halliwell's Film Guide 2006* gives a numeral followed by *m* after each film. That must be for the freaks, I thought, but soon I realized that *m* meant *minutes*, not *meter*. In scientific writing, *minutes* is abbreviated *min*; and *meter*, *m*.

General principles

You should introduce your abbreviations one by one as they first occur in the text, in this way:

nonsteroidal anti-inflammatory drug (NSAID).

But readers will miss this information if they turn directly from the abstract or introduction to the discussion section, as most readers probably do. Therefore it would be helpful if you also listed the abbreviations (headed: "Abbreviations used"). You could place this list either at the foot of the abstract page (see Chapter 23, "Typing") or on a following separate page. In the printed version, the list will usually appear as a footnote on the article's title page.

Finally, do not mix abbreviations and spelled-out terms; use either "nonsteroidal anti-inflammatory drug" or "NSAID" throughout the paper.

22

How to present statistical results

Too often, statistics are used "as a drunken man uses a lamp post, more for support than illumination" (Sumner 1992). Experts in the field can tell whether your study really needs statistics; if it does, they can help you to *plan* the statistical part of your study, for example, to estimate the sample size needed to demonstrate a difference (if it exists) and to choose appropriate statistical methods.

Then, when your study is completed, you will encounter another serious matter: how to *present* the statistical results. About half of such presentations contain statistical errors (Murray 1991). Here are the most common ones.

Using *mean* when *median* is meant

In a descriptive study on back pain in pregnancy, the women were asked to bend over with their arms hanging down. The distance between fingertips and floor was then measured. The result (mean and standard deviation) was reported as

12 ± 14 cm,

thus ranging between −2 and 26 cm, suggesting that some of the women must have poked their fingertips a couple of centimeters through the floorboards. This surprising conclusion is the result of reporting asymmetrically distributed (skewed) data by using *mean* (the average) and standard deviation instead of *median* (the value midway between the lowest and the highest value) and a percentile range, such as the interquartile range (25th to 75th percentile).

One rule of thumb says that if the standard deviation is greater than half the mean, the data are unlikely to be normally distributed (bell-shaped). In fact, most results in biomedical science are asymmetrically distributed (Lang and Secic 1997, 47).

If you present, in the same table, both normally and non-normally distributed data, this should be indicated in a footnote (see Chapter 8, "Typing the table").

Using *standard error* instead of *standard deviation*

Standard error of the mean (SEM; often incorrectly abbreviated as the unspecified SE) is invariably smaller than standard deviation (SD). It is therefore tempting to describe a set of observations with mean and SEM so as to suggest less variation in the observations. But to do so is inappropriate, as SEM, rather than being a descriptive term, reports the precision of an *estimate* of the mean in relation to its unknown value. SD, on the other hand, measures the spread of individual results around an *observed* mean.

Failure to distinguish between *statistical significance* and *biological importance*

In the following example, borrowed from Lang and Secic (1997, 58), a drug was found to lower the diastolic blood pressure by a mean of 8 mm Hg, from 100 to 92 mm Hg – which was statistically significant ($P < 0.05$).

However, as Lang and Secic say, a more informative way to estimate an effect is to construct a confidence interval. (Simply put, a 95 percent confidence interval [CI] is the range within which one can be 95 percent certain of including the true value.) In this case, a 95 percent CI was 2 to 14 mm Hg. This tells us that the reduction in blood pressure could be as much as 14 mm Hg, which would be *clinically* important, whereas a reduction of 2 mm Hg would not. Thus, a result can be statistically significant yet clinically inconclusive. In the running text the result could be presented in this way:

> Diastolic blood pressure was lowered by a mean of 8 mm Hg, from 100 to 92 mm Hg (95% CI = 2 to 14 mm Hg; $P = 0.02$).

Thus, P values estimate the statistical significance while confidence intervals also estimate the clinical significance. So, when the confidence interval is used, readers do not have to rely on the author's interpretation; they can judge for themselves.

Selected presentation of multiple statistical testing

Multiple testing can generate significant differences where none exist. With the conventional threshold of $P = 0.05$ to define a significant result, there is a 1 in 20 risk of finding a significant difference even when comparing two groups that are actually alike. To present only significant results of multiple testing, as if they were the only analyses performed, is inappropriate – to say the least. "If the fishing expedition catches a boot, the fishermen should throw it back, not claim that they were fishing for boots" (Mills 1993).

One way to accommodate the multiple testing problem is to adjust the P value by the Bonferroni method, that is, to divide the P value by the number of tests made. However, for large numbers of comparisons, the adjusted P value may be almost unattainable.

A better approach could be to decide, even in the planning stage, which test is of major interest and focus your attention on this variable when analyzing the data and writing the paper. Other data should be analyzed too, and interesting findings used for further research. (In fact, many fundamental breakthroughs stem from such unexpected findings.)

The next section deals with subgroup analyses, which pose problems similar to those of multiple testing.

Overinterpretation of subgroup effects

In a trial conducted on 16 027 patients with suspected acute myocardial infarction, Collins *et al.* (1987) made the incidental observation that the benefit of treatment was fourfold greater for patients born under the astrological sign of Scorpio than for patients born under all other signs put together. Computer searching through numerous subgroups makes it almost inevitable that some spurious "significant" results, like this one, will appear. However, it is reasonable to carry out a small number of subgroup analyses *– provided that these are specified in advance* (Altman 1995, 466). This kind of subgroup analysis could provoke ideas to be confirmed (or refuted) in future studies.

Another aspect of subgroup analysis deals with heterogeneity analysis in trials with statistically significant results, in order to assess whether the results are applicable to all patients (Rothwell 1995). For example, one study showed a statistically significantly higher perinatal mortality among newborns of immigrant women than among those of women of Swedish origin. Subgroup analyses, however, revealed that the results were applicable only to newborns of women from Sub-Saharan Africa, who had a particularly high perinatal mortality.

Using *relative* instead of *absolute* figures

In a large Swedish trial of mammography screening, a 24 percent reduction in mortality from breast cancer was reported. This impressive figure led to public clamor for screening programs, which would probably not have arisen if an *absolute* figure had been reported instead of a misleading *relative* figure. As breast cancer mortality declined from 0.51 to 0.39 percent, the reduction in absolute terms was actually only 0.12 percent.

A third way to present the result would be by giving the *Number Needed to Treat* in order to protect one of them from the disorder (Chatellier 1996). In this case, 833 women would have had to be screened regularly for 12 years to prevent a single death from breast cancer. This way of presenting results is easily understood by both doctors and patients. The relative risk reduction should therefore not be cited without simultaneously indicating the absolute risk reduction or Number Needed to Treat (Laupacis *et al.* 1992).

Finally, the results could be expressed also in terms of events per 100 000 of person-years.

Some further comments

The ± sign

The notation of an observed mean as 12.3 ± 0.4 gives no indication as to whether the second figure is a standard deviation or something else. A clearer presentation would be:

> the mean was 12.3 (SD 0.4)

or:

> the mean (SD) was 12.3 (0.4).

With this construction you also avoid the ± sign. Some journals do not allow its use.

P < 0.05 ≠ the truth

A firmly ingrained idea is that $P < 0.05$ = the truth, while $P > 0.05$ = unpublishable. But P values of 0.04 and 0.06, which differ very little, ought to lead to similar interpretations rather than radically different ones. To emphasize this point, the editor of one of the world's major journals, *The Lancet*, was prepared to strip! At a workshop for editors, he removed his jacket, tie, and shirt to display a T-shirt bearing a crossed-out sign stating $P < 0.05$ (Crossan and Smith 1996). Some journals now ask for exact P values when values fall above 0.001. The sign < (less than) attached to the P would thus be used only at the extreme $P < 0.001$.

23

Typing

On the following pages I will show you a manuscript with a layout that can almost invariably be used as a model when typing your own. It follows the Vancouver recommendations but, as a model, can be adapted to comply with most other instructions.

Figure 23.1 The editor at work. (Cartoon by Louis Hellman, first published in H.E. Emson. *BMJ* 1994; **309**:1738; reproduced with permission.)

How to Write and Illustrate a Scientific Paper

If you follow this layout, you will never again need to ask yourself in front of your word processor: "Well, how shall I do it this time?" Instead, you can concentrate on what you have to say. Again I have used *Once upon a time* as the running text.

Text in 12-point Times and **double-spaced**

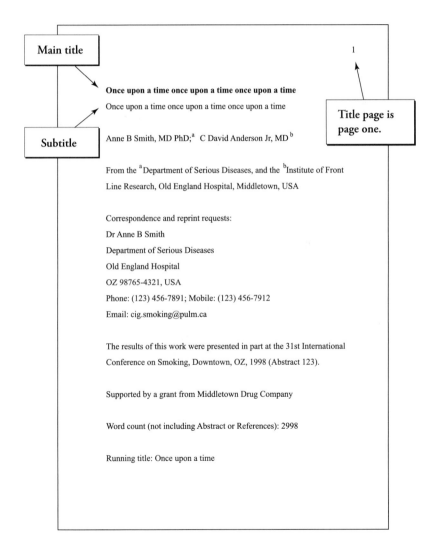

ABSTRACT

Background. Once upon a time once upon a time once upon a time. Once upon a time once upon a time.

Methods. Once upon a time. Once upon once a time once upon a time once upon a time. Once upon a time once upon a time.

Results. Once upon a time once upon a time once upon a time. Once upon a time once upon a time once upon a time. Once upon a time.

Conclusions. Once upon a time.

> **Key words, written in alphabetical order, to be chosen from the list of Medical Subject Headings (MeSH) of *Index Medicus*. Available from: www.nlm.nih.gov**

Key words. A time; Once;

Abbreviations used. OU, once upon; OUT, once upon a time; UT, upon a time.

> **Abbreviations listed alphabetically; also explained in text at first mention.**

If possible, use no more than three levels of headings, emphasized by lettering in:

1. **BOLD (HEAVY) CAPITALS,**

2. **bold lowercase,**

3. *italics (cursive) style.*

This layout makes it possible to distinguish the headings using the 12-point Times letter size throughout.

METHODS

Patients

Recruitment

Once upon a time once upon a time once upon a time once upon a time once upon a time. Once upon a time. Once upon a time once upon a time. Once upon a time. Once upon a time. Once upon a time once upon a time. Once upon a time

once upon a time.

Three "returns" when a heading follows

Diagnostic criteria

Two returns after a heading

Once upon a time once upon a time. Once upon a time once upon a time once upon a time. Once upon a time once upon a time. Once upon a time. Once upon a time once upon a time. Once upon a time once upon a time once upon a time. Once upon a time.

Two returns between paragraphs; no indention needed

Once upon a time once upon a time. Once upon a time once upon a time once upon a time. Once upon a time once upon a time. Once upon a time once upon a time.

Treatment

Once upon a time once upon a time once upon a time once upon a time once upon a time. Once upon a time. Once upon a time once upon a time. Once upon a time. Once upon a time. Once upon a time once upon a time. Once upon a time once upon a time. Once upon a time once upon a time. Once upon a time once upon a time.

Laboratory methods

Once upon a time once upon a time. Once upon a time once upon a time once upon a time (Figure 1). Once upon a time. Once upon a time. Once upon a time. Once upon a time once upon a time once upon a time once upon a time once upon a time. Once upon a time. Once upon a time once upon a time once upon a time.

[Place Figure 1 about here]

Statistical analysis

Once upon a time once upon a time (Table 1). Once upon a time once upon a time once upon a time. Once upon a time once upon a time

[Table 1 about here]

RESULTS

Patient characteristics

Once upon a time once upon a time once upon a time once upon a time once upon a time. Once upon a time. Once upon a time once u-pon a time. Once upon a time. Once upon a time. Once upon a time once upon a time. Once upon a time once upon a time. Once upon a time once upon a time. Once upon a time once upon a time. Once upon a time once upon a time.

> The most common error in submitted manuscripts is single spacing or one-and-a-half line spacing; insertion of editorial alterations requires at least double spacing.

Outcome of the study

> A few journals require triple spacing.

Once upon a time once upon a time. Once upon a time once upon a

time once upon a time. Once upon a time. Once upon a time. Once upon

a time. Once upon a time once upon a time.

Adverse effects

> Most journals, however, ask for double spacing.

Once upon a time once upon a time once upon a time once upon a

time once upon a time. Once upon a time. Once upon a time once u-

pon a time. Once upon a time. Once upon a time. Once upon a time

once upon a time. Once upon a time once upon a time. Once upon a

time once upon a time. Once upon a time once upon a time. Once

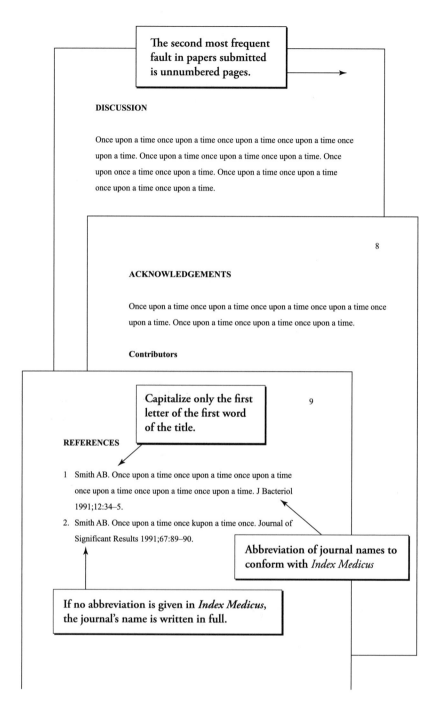

The second most frequent
fault in papers submitted
is unnumbered pages.

DISCUSSION

Once upon a time once upon a time once upon a time once upon a time once
upon a time. Once upon a time once upon a time once upon a time. Once
upon once a time once upon a time. Once upon a time once upon a time
once upon a time once upon a time.

8

ACKNOWLEDGEMENTS

Once upon a time once upon a time once upon a time once upon a time once
upon a time. Once upon a time once upon a time once upon a time.

Contributors

Capitalize only the first
letter of the first word
of the title.

9

REFERENCES

1 Smith AB. Once upon a time once upon a time once upon a time
   once upon a time once upon a time once upon a time. J Bacteriol
   1991;12:34–5.
2. Smith AB. Once upon a time once kupon a time once. Journal of
   Significant Results 1991;67:89–90.

Abbreviation of journal names to
conform with *Index Medicus*

If no abbreviation is given in *Index Medicus*,
the journal's name is written in full.

How to Write and Illustrate a Scientific Paper

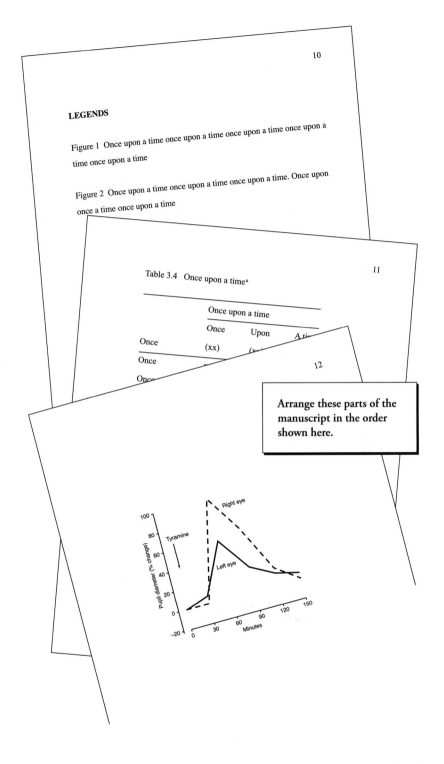

LEGENDS

Figure 1 Once upon a time once upon a time once upon a time once upon a time once upon a time

Figure 2 Once upon a time once upon a time once upon a time. Once upon once a time once upon a time

Table 3.4   Once upon a time[a]

	Once upon a time		
	Once	Upon	A ti
Once	(xx)	(x	
Once			
Once			

Arrange these parts of the manuscript in the order shown here.

"Twain spacing"

Samuel L. Clemens (Mark Twain) claims in his autobiography that *Tom Sawyer* (published in 1876) was the first typewritten book manuscript (White 1988). It was double-spaced.

Although most authors now double-space the main text of their manuscripts, many appear to regard single spacing as adequate for the reference list. Yet this is the section in which most editorial changes have to be made. Single-spaced text is impossible to edit clearly.

The sole exception from double spacing is text within a table that forms groups of words, where each group may be single-spaced, with a double space between groups (see also Chapter 8, "Typing the table"):

Once upon a time once upon a time once upon a time	12
Once upon a time once upon a time once upon a time	34

12-point Times

The Times font is a standard choice for newspapers and other periodicals. (It was cut originally for the daily newspaper *The Times* of London.) Text set in Times is easy to read and its compact design saves space. So, Times might be the right choice for you when typing the manuscript. A font size of 12 points is recommended.

There is one exception, however. Helvetica, a sans serif typeface, is considered better for text that is not intended for continuous reading, such as that of graphs.

Avoid using block capitals

TEXT SET IN CAPITALS (BIG LETTERS) IS MORE DIFFI-CULT TO READ AS CAPITALS HAVE NO DISTINCT SHAPE, whereas lowercase letters (small letters) have ascenders, such as *b*, and descenders, such as *p*, which distinguish the letters from each other and make reading easier.

However, in first-level headings, such as RESULTS, capitals are recommended, in order to distinguish these headings from sub-headings.

Up-and-down-style versus down style

Most journals in the USA use the "up-and-down-style" in titles and headings. This style means that the first letter of each word is capitalized. Exceptions are articles (a, the), prepositions (on, in), and coordinate conjunctions (and, or). Here is an example from *N. Engl. J. Med.* (Grüters *et al.* 1995):

> Persistence of Differences in Iodine Status in Newborns after
> the Reunification of Berlin.

Most other countries use the "down style," also called sentence style, meaning that only the first letter of the first word is capitalized:

> Persistence of differences in iodine status in newborns after the
> reunification of Berlin.

In this book I have used the down style, except for the title of the book (*How to Write and Illustrate a Scientific Paper*). Well, I have never said that this style is without charm, only that down style is more reader-friendly – at least for an international readership.

Do not mimic the journal's style

Many of the 600-odd journal publishers that have adopted the Vancouver style do not themselves adhere to its requirements in every detail. They do recommend, however, that authors submitting manuscripts "should not try to prepare them in accordance with the [journal's style] but should follow the [Vancouver requirements]" (www.icmje.org; accessed 20 January 2002). The purpose of this recommendation is to relieve authors of time-consuming and unproductive effort to make their manuscripts comply with any one journal's particular requirements. Any necessary changes will be made by the journal's copy editor.

The length of the manuscript

Editors are biased in favor of short articles. Even hard facts can be explained in a few pages. A classic example is the report by Watson and Crick (1953) on the structure of DNA, which occupies just over one page in *Nature*, and is understandable even to non-chemists.

Most biomedical journals will not accept manuscripts exceeding 3000 to 4000 words (plus references, figures, and tables). As one double-spaced page takes about 300 words, that means a maximum of about 10 to 14 pages of running text. However, presentation of results from certain subjects, such as occupational science, medical ethics, and nursing and health care, may need more space. Check the Instructions for Authors of your target journal for the number of words allowed. And never exceed that number! Never!!

Let us postulate that yours is a 12-page paper. To achieve fair proportions between the different parts of your paper, you should devote about one page to the introduction, and about 3–4 pages each to the methods, results, and discussion sections.

The importance of punctuation

Pay attention to punctuation. For example, a comma placed incorrectly can make a difference, as in Lynne Truss's *Eats, Shoots & Leaves* (2003). As Truss explains on the jacket, a panda walks into a café, eats a sandwich and shoots a gun into the air. On his way out, he tosses a badly punctuated wildlife manual at the confused waiter and tells him to turn to the section about his species. The waiter turns to the page and reads:

> *Panda.* Large black-and-white bear-like mammal, native to China. Eats, shoots and leaves.

I saw the following passage in the weekly magazine *Time*. In the absence of appropriate punctuation, it will be misunderstood:

> Woman without her man has no reason for living,

which should read:

> Woman: without her, man has no reason for living.

Some learn the typographical conventions early, like the five-year-old girl who was asked to explain why we have "dots" at the end of sentences (Henshaw, quoted by Hartley 1994). She answered:

> It's to finish a sentence. If you don't put a dot and you write a letter people might think you've forgotten to post the other half.

24

Dealing with editors and referees

Here is a question from a course participant:

> Am I entirely left to the tender mercies of the editors and the referees? Or do I dare to argue for my own view when I feel that the referee might have misunderstood a certain point? Am I impolite if I do so?

No, you aren't – if you do it politely. Thus not exactly in the way quoted below from a covering letter to the editor of *Cardiovascular Research* (Hearse and the Editorial Team 1992):

> Many of the "problems" the referee had with our manuscript appear to stem from his limited understanding of electrophysiology or from our failure to explain observations at a more basic level.

In this case the referee happened to be a most eminent researcher in electrophysiology. Try instead to write as though the referee were God the Father Himself. But don't hesitate to make your point:

> Thank you for the constructive criticism of my paper. Here are my comments on the referee's suggestions.

Page 3, lines 2–5. What I wanted to say here was . . .
I have rewritten this passage to make my point of view more clear.
Page 4, lines 3–5. . . .

Don't forget that the referee might have sacrificed hours of unpaid effort on your manuscript.

Do referees delay?

Here is another question from a course participant:

> How big is the risk that the paper goes to a competitor who delays the whole thing?

That referees delay publication while they incorporate the ideas of the refereed work into their own publication is extremely rare. With few exceptions, referees are honorable men and women.

If these words do not allay your fears, what can you do to protect your ideas from being stolen? One way could be to present your results at a conference before submitting the paper for publication – your results will then be safeguarded in a conference abstract. But can you be absolutely sure that the person reading your submitted abstract will not leak your ideas before the abstract is published? No, there can be no such certainty. So, if you cannot live with that, you might consider not publishing – and science will be deprived of your interesting findings.

Unpublished work

The referee (also known as reviewer) must have access to all papers you refer to in your manuscript. So when you submit your manuscript, enclose copies of any works "in press," "in manuscript," or "in preparation" that you mention in your paper. If you haven't done so, the assessment of your work may be incomplete. One reviewer wrote:

> Much of the key cited methodological material is "in press" and cannot be judged by this reviewer.

But don't forget that referees have access to their own works. One referee told me in her answer:

> A lot of what they say is virtually a direct quotation of my own paper.

Shortening the manuscript

An author had been asked to shorten the text of his manuscript. In his (serious?) covering letter, he said (Baumeister 1992):

> You suggested that we shorten the manuscript by 5 pages, and we were able to accomplish this very effectively by altering the margins and printing the paper in a different font with a smaller type face. We agree with you that the paper is much better this way.

A more ambitious author did it in this way:

> As suggested in your letter, we have reduced the text by close to 30%. The word count in the revised version, compared with the previous version:
>
> | | Is now | Was | Percent cut |
> |---|---|---|---|
> | Introduction | 386 | 520 | 26 |
> | Methods | 1006 | 1605 | 37 |
> | Results | 2037 | 2762 | 26 |
> | Discussion | 1182 | 1561 | 20 |
> | Total count | 4611 | 6345 | 27 |

Perhaps you can find a middle-of-the-road approach.

Figure 24.1 The editor at his desk.

Accepted or rejected

If your paper is accepted, you may receive a preprinted card with a short statement, as I did:

> Dear Doctor:
> Your manuscript has been accepted for publication. It is now being sent to the Publisher and in due time you will receive a proof.

In contrast, a letter of rejection may seem almost movingly considerate, as this one I received from *The Lancet:*

> Dear Dr Gustavii,
> I hope we shan't dismay you by failing to accept this paper. I should like to have enabled our readers to see your further interesting findings but at present we are in such trouble from pressure on space. . . . I am sure you will readily make an alternative arrangement.

I was dismayed, but found consolation in the fact that I was not alone. In 1937, *Nature* rejected a submission from Hans Krebs in which he described the citric acid cycle – one of the central features of cellular metabolism, now known as the Krebs cycle. The paper was accepted instead by the editor of *Enzymologia*. In 1953, Krebs was awarded a Nobel Prize in recognition of his work.

25

Correcting proofs

Popeye, the beloved cartoon character, would probably never have been created had it not been for a misplaced decimal point. As you know, Popeye gets his strength by eating spinach, assumed to be rich in iron. This misconception derives from a report indicating, due to a misplaced decimal point, that spinach has an iron content tenfold higher than its true value. An overlooked error seldom has such amusing consequences, however.

How to read proof

When you receive your masterpiece, nicely typeset in the form of a proof, you may be tempted to read it straight through at that very moment. My advice is to follow your intuition. You will be on the alert and will easily notice if the reading makes sense, thus catching errors of omission, such as a dropped line or a lost paragraph. In order not to overlook printer's errors, however, you will have to reread the proof at least once more.

For the second reading, persuade someone to slowly read the manuscript aloud while you check the text in the proof. If you can't find a reader, place a finger under the first line of the manuscript and a finger under the first line of the proof, just under the first character. Look from manuscript to proof and back again, checking word by word, numeral by numeral, and punctuation mark by punctuation mark. Be especially careful in checking the tables and the reference list.

Another option that I have heard of but not used is to read the manuscript into a tape recorder and then listen to the recording while looking at the proof.

If you check the proof too hastily, you may live to regret it. I once overlooked a mistyped numeral and afterward had to correct it by hand in 300 reprints.

What to correct

The main reason for sending you the proofs is so you may correct the typing errors. At this stage, you are not allowed to polish the prose. You may, however, correct a serious mistake, such as inconsistency between data in the abstract and in the main text of the paper.

Moreover, if you have changed your mind about part of the content, or if you have acquired relevant new information, you can write an addendum (also called "Notes added in proof") placed at the end of the main text, before the list of references. The main reason for using an addendum is the ethical aspect of adding to the body of the text some new matter or a revision containing material not seen by the referee. Here is an example of presentation of new information (Federle *et al.* 1982):

> **Addendum**
> Since this manuscript was submitted, 25 additional patients have been studied. DR [Digital Radiography] pelvimetry has completely replaced the conventional method at this institution.

If you have referred to a paper as in "in press" and this paper has already been published, update it by providing the volume, year, and pagination.

Correction marks

There are several systems of proof correction: the continental European system, the British system, the American system, and various systems used in other countries. But typesetters worth their salt can cope with them all. So there is no need for you to learn more than one system. And the one I recommend is the American system, as, in my opinion, it is the easiest to use. On the following pages you are shown the most common American proofreader's marks and how to use them.

Unlike corrections in the manuscript, corrections in the proof must be marked twice, once at the point where the error occurs and once in the margin. Typesetters scan the margins and won't notice corrections indicated in the text if you have failed to indicate the change in the margin.

You need not know more than the marks given in this chapter; they cover most cases of correction in the average proof. But, in those few cases where you cannot find a suitable mark here, just write an instruction to the typesetter in the margin and circle it. Write the correction beside the instruction and make an appropriate mark in the text to show where the new, corrected material is to be inserted.

Electronic proof

Technology now exists whereby those correcting electronic proofs can electronically show correction.

Table 25.1 Commonly used American correction marks

Mark the text	Meaning	In the margin
On͡ce upon	Close up	◡
Once|upon	Add space	#
upon a ∧ time	Equalize spacing	eq #
A time∂	Delete (take out)	ℛ
a ti/me	Delete and close up	ℛ
a t/yme	Substitute	i
there ~~were~~	Substitute	was
Once, but not twice ∧	Insert period	⊙
Once ∧ but not twice.	Insert comma	⋏
⌐upon|Once|	Transpose	(tr)
once̳	Capitalize	(cap)
Once ɤpon	Change to lower case (small letter)	(lc)
The Sleeping Beauty	Set in italic (cursive)	(ital)

How to Write and Illustrate a Scientific Paper

Here is a sample of text with errors marked, corrected, and commented upon.

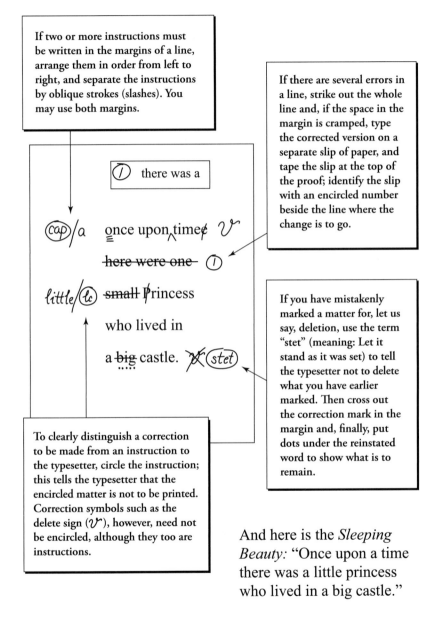

If two or more instructions must be written in the margins of a line, arrange them in order from left to right, and separate the instructions by oblique strokes (slashes). You may use both margins.

If there are several errors in a line, strike out the whole line and, if the space in the margin is cramped, type the corrected version on a separate slip of paper, and tape the slip at the top of the proof; identify the slip with an encircled number beside the line where the change is to go.

To clearly distinguish a correction to be made from an instruction to the typesetter, circle the instruction; this tells the typesetter that the encircled matter is not to be printed. Correction symbols such as the delete sign ($\mathcal{V}$), however, need not be encircled, although they too are instructions.

If you have mistakenly marked a matter for, let us say, deletion, use the term "stet" (meaning: Let it stand as it was set) to tell the typesetter not to delete what you have earlier marked. Then cross out the correction mark in the margin and, finally, put dots under the reinstated word to show what is to remain.

And here is the *Sleeping Beauty:* "Once upon a time there was a little princess who lived in a big castle."

26

Authors' responsibilities

Subjects' right to privacy

How to protect a subject's identity? In this figure, executed with a dash of humor, the anonymity of both subjects has been protected by the traditional black band across the eyes.

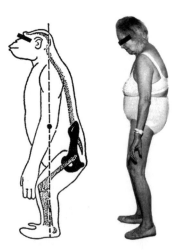

Figure 26.1 Simian stance, which resembles the posture of an anthropoid ape, can be a valuable clue to diagnosis of spinal stenosis. (Reproduced, with permission, from P.A. Simkin. Simian stance: a sign of spinal stenosis, *The Lancet* 1982; **ii (8299)**:652–3, © The Lancet Ltd.)

How to Write and Illustrate a Scientific Paper

Nevertheless, a few weeks later, in *Letters to the Editor*, a reader told us that despite the bar across the animal's eyes, he had immediately recognized it as one he had seen in Melbourne Zoo and suggested that authors should take greater care to preserve anonymity when presenting ape data (Millar 1982).

This observation tells us that a black band across the eyes may be insufficient to disguise the subject. So, in cases like this, informed consent should be obtained, as recommended by the International Committee of Medical Journal Editors (1995). The woman in the figure had given such consent. Simkin's figure also tells us that authors need not always be deadly serious in their reporting. Rather, humor can help to convey the message.

Duplicate submission

One of my course participants asked:

> If I submit a paper to a journal, can I at the same time, in a revised form, send a paper (same subject, same material) to another journal? Or would my young research career finish there and then?

Yes, that would be the end! A unanimous jury would give you nothing less than a life sentence for self-plagiarism, as happened to an author who had submitted the same manuscript to two journals in the USA. All American journals publishing in that field agreed never again to consider any manuscript from that scientist's laboratory (Abelson 1982).

Borrowing published material

Here is another question from a course participant:

> If I redraw a published picture and make some small alterations, can I then call it "my own"?

No, usually not. If the published picture has "original features," and if these are retained in the redrawn figure, then you should seek permission to reproduce the picture. Now, what constitutes the originality of an illustration? As even judges in lawsuit cases on this subject can disagree, we can do no better than use our intuition to find an answer when this question is raised. Here is my own personal view on two cases.

I saw the original Viking cartoon in a daily newspaper, (top left in Figure 26.2). The Viking encourages the readers to say NEJ (NO) to the European Union, a subject far from teaching how to write a scientific paper. But I found the figure useful for the purpose by replacing the text on the stone and, after whiting out the eye and mouth regions, remaking the facial expression. However, these alterations do not make the redrawn figure "mine," as its original features are retained almost unchanged. I therefore had to obtain permission to publish it (bottom in Figure 26.2; see also Figure 2.1).

Figure 26.2 Permission is needed for publication of the redrawn cartoon (*bottom*), as the originality of the picture (*top left*) is retained. (Redrawn, with permission, from Majewski 1994.)

How to Write and Illustrate a Scientific Paper

On the other hand, when a figure's characteristics are not copied, permission is not needed. For example, I used the textbook figure (the left one in Figure 26.3) without permission as a model when drawing the figure to the right.

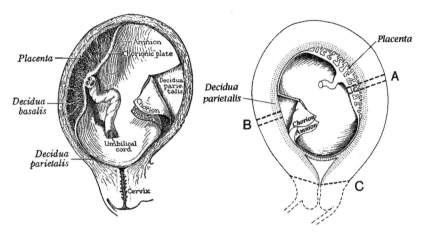

Figure 26.3 Permission not needed for the redrawn figure (right), as the characteristics of the original figure (left) are not copied.

However, to reproduce these two figures in the present book, I did have to obtain permission. Thus, the left figure is reproduced, with permission, from Arey 1954; that on the right is reproduced from Gustavii 1975, with the publisher's permission. Note that I had to obtain permission to reproduce even a figure I myself had drawn for a paper I had authored myself. As is often the case, I was required before the publication of my paper to sign a statement transferring my copyright to the publisher, including both text and illustrations.

When you seek permission to reproduce a figure (original or redrawn), write not only to the publisher but also to the author and, if appropriate, to the illustrator as well. For example, the copyright owner of Figure 23.1 in this book appeared to be the artist. (The publisher told me so and gave me her address.) Thus, although the copyright often belongs to the publisher, this is not always the case.

Most journals require permission from both copyright owner and author. Permission is needed also to reproduce a table (or part of it) or to quote text at length (say, 100 words or more).

Many publishers give permission freely, but some charge a fee. For the copyrighted material (44 graphs, seven tables) in the first edition of this book, fees were charged for three graphs and one table; the graphs cost me (in 1998) £30, £30, and US$20; the table, US$25. Most publishers responded to my request for permission within one to three months. One publisher, however, replied only after ten months and two reminders.

Nowadays, however, many publishers of scientific journals provide online forms for requesting permission. Because these forms can be exhaustingly detailed, it may be easier for you to send a request by e-mail. Use a template (Figure 26.4) that you can save in your computer. Permission is often granted within hours, even minutes – if the request is received during working-hours.

The old template in Figure 26.5 is still useful. For example, in 2006, when I sought permission from a number of authors, I was unable to find e-mail addresses in two cases, and I had to send the template by regular mail.

The wording of the credit line I suggest, for example, "Reproduced, with permission, from Simkin 1982," is not always accepted. Some copyright holders wish to specify in detail the wording of the acknowledgment, which must be followed strictly, even if it means that the credit line becomes much longer than the figure legend (see Figure 14.1).

27 July 2006

Catherine Nielsen
Copyright Manager
Elsevier
Health Sciences Rights Department
1600 John F. Kennedy Boulevard, Suite 1800
Philadelphia, PA 19103-2899, USA

Dear Ms Nielsen,

I am preparing a 2nd edition of my book *How to Write and Illustrate a Scientific Paper,* Cambridge University Press. I would greatly appreciate your permission to reproduce figure 2 from:

> Chaparro, C. M.; Neufeld, L. M.; Alavez, G. T.; Cedillo, R. E-L.; Dewey, K.G. 2006. Effect of timing of umbilical cord clamping on iron status in Mexican infants: a randomised controlled trial. *The Lancet* 367:1981–9.

This figure would help authors to design a box plot.

Acknowledgments of the source will be printed on the page where the figure appears, as follows:

> Reproduced, with permission, from Chaparro et al. 2006.

A full reference will be given in the reference list. If this form of acknowledgement is not sufficient, please indicate how the credit line should appear.

Yours sincerely,

```
Björn Gustavii, MD, PhD
Clemenstorget 3
SE-22221 LUND, Sweden
bjorngustavii@telia.com
```

Figure 26.4 A template to be used when requesting, by e-mail, permission to use copy-righted material.

Avdelningen för Obstetrik och Gynekologi
Universitetssjukhuset i Lund

Department of Obstetrics and Gynecology
Lund University Hospital

23 February 1998

The Publisher of *The Lancet*
42 Bedford Square
London WC1B 3SL, UK

To the Publisher:

I am preparing a book tentatively entitled *Scientific Writing*. I would greatly appreciate your permission to include the figure (see enclosure) from:

> Simkin PA. Simian stance: a sign of spinal stenosis. Lancet 1982;ii:652-3.

Acknowledgment of the source will be printed on the page where the figure appears, as follows:

> Reproduced, with permission, from Simkin 1982.

A full reference will appear in the reference list. If this form of acknowledgment is not sufficient, please indicate below how the credit line should appear.

I am also writing to Professor Simkin requesting permission to reproduce this figure.

If you are willing to let me use the figure, please sign and return this letter (a copy of the letter is enclosed for your records).

Many thanks for your help.

Yours sincerely,

Björn Gustavii, MD, PhD

Permission granted by the owner, The Lancet Ltd. provided complete credit is given to the original source and © owner. The credit line should include the name of the author(s), title of the article, volume #, issue #, inclusive pages, © by The Lancet Ltd. and Year of publication.

I grant permission to use the figure in your book. 11/3/98.

Date: ....

Signed: ....Joanne Cuill

Credit line to be used:

Address:
Department of Obstetrics and Gynecology
University Hospital
SE-221 85 LUND
SWEDEN

Telephone:
(46) 46-17 10 00

Telefax:
(46) 46-15 78 68

Figure 26.5 A request sent by regular post.

Saving your original data

When you have completed your paper you may wonder what to do with all the raw material you have collected. Should you discard it? No, definitely not. Your target journal may require access to it. For example, if you submit a paper to *JAMA*, you have to sign a statement saying, "I certify that if requested, I will provide the data or will cooperate fully in obtaining and providing the data on which the manuscript is based for examination by the editors or their assignees" (*JAMA* 2002). Even after publication, you are obliged to provide those who request it, with the original data of your paper – a responsibility you will bear for at least five years after publication (some say 10 years).

Literature needed on your desk

On phraseology

Hornby, A. S.; Ashby, M. 2005. *Oxford Advanced Learner's Dictionary of Current English*. 7th edn. New York: Oxford University Press.

On synonyms

The Merriam-Webster Dictionary of Synonyms and Antonyms. 1998. Springfield, MA: Merriam-Webster.

On manuscript preparation

International Committee of Medical Journal Editors. *Uniform Requirements for Manuscripts Submitted to Biomedical Journals* (the Vancouver Document, www.icmje.org).

On how to abbreviate a journal's title

List of journals in the *Index Medicus*. Bethesda, MD: National
Library of Medicine; published annually as a list in the
January issue of *Index Medicus*. For sale on the Net by
amazon.com, as a separate publication. Also available at:
www.nlm.nih.gov

Further reading

Guides to writing

Albert, T. 2000. *The A–Z of Medical Writing.* London: BMJ Books.

Alley, M. 1997. *The Craft of Scientific Writing.* 3rd edn. New York: Springer-Verlag.

Booth, V. 1993. *Communicating in Science. Writing a Scientific Paper and Speaking at Scientific Meetings.* 2nd edn. Cambridge: Cambridge University Press.

Browner, W. S. 1999. *Publishing and Presenting Clinical Research.* Baltimore: Williams & Wilkins.

Byrne, D. W. 1998. *Publishing Your Medical Research Paper. What They Don't Teach You in Medical School.* Baltimore: Williams & Wilkins.

Davis, M. 2004. *Scientific Papers and Presentations.* 2nd edn. San Diego: Academic Press.

Day, R. A.; Gastel, B. 2006. *How to Write and Publish a Scientific Paper.* 6th edn. Westport, CT: Greenwood Press.

Goodman, N. W.; Edwards, M. B.; Black, A. 2006. *Medical Writing: A Prescription for Clarity.* 3rd edn. Cambridge: Cambridge University Press.

Hall, G. M., editor. 2003. *How to Write a Paper.* 3rd edn. London: BMJ Books.

How to Write and Illustrate a Scientific Paper

Huth, E. J. 1999. *Writing and Publishing in Medicine.* 3rd edn. Baltimore: Williams & Wilkins.

Iles, R. L. 2003. *Guidebook to Better Medical Writing.* Washington, D.C.: Island Press.

King, S. 2002. *On Writing. A Memoir of the Craft,* pp. 139–288. New York: Pocket Books.

Malmfors, B.; Garnsworthy, P.; Grossman, M. 2005. *Writing and Presenting Scientific Papers.* 2nd edn. Nottingham: Nottingham University Press.

Matthews, J. R.; Bowen, J. M.; Matthes, R. W. 2001. *Successful Scientific Writing. A Step-by-Step Guide for the Biological and Medical Sciences.* 2nd edn. Cambridge: Cambridge University Press.

Montgomery, S. L. 2002. *The Chicago Guide to Communicating Science (Chicago Guides to Writing, Editing, and Publishing).* Chicago: University of Chicago Press.

O'Connor, M. 1991. Reprint 1999. *Writing Successfully in Science.* London: E & FN Spon.

Schoenfeld, R. 1989. *The Chemist's English.* 3rd edn. New York: Wiley-VCH.

Woodford, F. P. 1999. *How to Teach Scientific Communication.* Bethesda: Council of Biology Editors.

Taylor, R. B. 2005. *The Clinician's Guide to Medical Writing.* New York: Springer.

Zeiger, M. 1999. *Essentials of Writing Biomedical Research Papers.* 2nd edn. New York: McGraw-Hill Professional Publishing.

Punctuation

Carey, G. V. 1976. *Mind the Stop. A Brief Guide to Punctuation.* London: Penguin.

Truss, L. 2003. *Eats, Shoots & Leaves. The Zero Tolerance Approach to Punctuation.* London: Profile Books.

The English language

Day, R. A. 1995. *Scientific English. A Guide for Scientists and Other Professionals.* 2nd edn. Phoenix: Oryx Press.

Strunk, W. Jr.; White, E.B.; Angell. R. 2000. *The Elements of Style.* 4th edn. Boston: Allyn & Bacon.

Style manuals

The ACS Style Guide: Effective Communication of Scientific Information (An American Chemical Society Publication). 2006. 3rd edn. Edited by Anne M. Coghill and Lorrin R. Garson. Washington, DC: American Chemical Society.

American Medical Association Manual of Style. 2007. 10th edn. New York: Oxford University Press.

Publication Manual of the American Psychological Association. 2001. 5th edn. Washington, DC: American Psychological Association.

Science & Technical Writing. A Manual of Style. 2001. 2nd edn. Edited by Philip Rubens. New York: Routledge.

Scientific Style and Format: The CBE Manual for Authors, Editors, and Publishers. 1994. 6th edn. Edited by Edward J. Huth. New York: Cambridge University Press.

Illustrations

Briscoe, M. H. 1996. *Preparing Scientific Illustration. A Guide to Better Posters, Presentations, and Publications.* 2nd edn. New York: Springer-Verlag.

Harris, R. L. 1999. *Information Graphics. A Comprehensive Illustrated Reference.* New York: Management Graphics.

Tufte, E. R. 2001. *The Visual Display of Quantitative Information.* Cheshire, CT: Graphics Press.

How to Write and Illustrate a Scientific Paper

Statistics

Altman, D. G. 2006. *Practical Statistics for Medical Research.* 2nd edn. London: Chapman & Hall.

Lang, T. A.; Secic, M. 2006. *How to Report Statistics in Medicine. Annotated Guidelines for Authors, Editors, and Reviewers.* 2nd edn. Philadelphia: American College of Physicians.

Literature cited

Abelson, P. H. 1982. Excessive zeal to publish. *Science* **218**:953.

Altman, D. G. 1995. *Practical Statistics for Medical Research.* London: Chapman & Hall.

Altman, D. G. 1985. Comparability of randomised groups. *Statistician* **34**:125–36.

Andersson Grönlund, M. 2005. Ophthalmologic characteristics and neuropediatric findings – with special emphasis on children adopted from Eastern Europe. (Dissertation.) Göteborg, Sweden: The Sahlgrenska Academy at Göteborg University.

Andolf E. 1989. Sonography of the female pelvis with emphasis on ovarian tumours. (Dissertation.) Lund, Sweden: University of Lund.

Animal Info. 2002. Information on rare, threatened and endangered mammals. Severna Park (MD): Animal info. www.animalinfo.org (accessed 2 February 2002).

Arey, L. B. 1954. *Developmental Anatomy. A Textbook and Laboratory Manual of Embryology,* p. 130. 6th edn. Philadelphia: Saunders.

Azem, J. 2005. Approaches to analyses of cytotoxic cells, and studies of their role in *H. pylori* infection. (Dissertation.) Göteborg, Sweden: The Sahlgrenska Academy at Göteborg University.

Bailar, J. C. 1986. Science, statistics, and deception. *Ann. Intern. Med.* **104**:259–60.

Baker, J. R. 1955. English style in scientific papers. *Nature* **176**:851–2.

Baranto, A. 2005. Traumatic high-load injuries in the adolescent spine. Clinical, radiological and experimental studies. (Dissertation.) Göteborg, Sweden: The Sahlgrenska Academy at Göteborg University.

How to Write and Illustrate a Scientific Paper

Baumeister, R. F. Dear Journal Editor, it's me again. Dialogue [cited by Hearse, 1992].

Begg, C.; Cho, M.; Eastwood, S.; Horton, R.; Moher, D.; Olkin, I.; *et al.* Improving the quality of reporting of randomized controlled trials. The CONSORT statement. *JAMA* 1996; **276**:637–9.

Begley, S. 1996. To stand and raise a glass. *Newsweek,* July 1, 42–5.

Bengtsson, L. P. 1968. Therapeutic abortion by means of intra-uterine injections. Techniques, effects, risk and mechanisms of effect. *Medical Gynecology & Sociology 3:6–14.*

Bennet, S. 1992. *Churchill* [documentary]. London: British Broadcasting Corporation.

Bergström, A. 1994. Experimental retinal cell transplants. (Dissertation.) Lund, Sweden: University of Lund.

BMJ editorial 1985. Reference 13. **291**:1746.

Broad, W. J. 1981.The publishing game: Getting more for less. Meet the least publishable unit, one way of squeezing more papers out of a research project. *Science* **211**:1137–9.

Brownlow, K.; Gill, D. 1983. *Unknown Chaplin* [video]. Thames Video Collection.

Burrough-Boenisch, J. 2006. EATAW (European Association for the Teaching of Academic Writing) email forum. www.eataw.org/listserv/ (accessed 8 February 2006).

Carling, P. 2006. EATAW (European Association for the Teaching of Academic Writing) email forum. www.eataw.org/listserv/ (accessed 9 December 2006).

CBE. 1994. *See* Council of Biology Editors' Style Manual Committee.

Chaparro, C. M.; Neufeld, L. M.; Alavez, G. T.; Cedillo, R. E-L.; Dewey, K.G. 2006. Effect of timing of umbilical cord clamping on iron status in Mexican infants: A randomised controlled trial. *The Lancet* **367**:1981–9.

Chaplin, C. 1973. *My Autobiography*, p. 208. Harmondsworth: Penguin.

Chapman, M.; Mahon, B. 1986. *Plain Figures*, p. 74. London: HMSO.

Chatellier, G.; Zapletal, E.; Lemaitre, D.; Menard, J.; Degoulet, P. 1996. The number needed to treat: A clinically useful nomogram in its proper context. *BMJ* **312**:426–9.

Chernin, E. 1988. The "Harvard system": A mystery dispelled. *BMJ* **297**:1062–3.

Christensen, K. K. 1980. Group B Streptococci. Aspects on urogenital epidemiology and obstetrical significance. (Dissertation.) Lund, Sweden: University of Lund.

Clarke, K. W.; Gray, D.; Keating, N. A.; Hampton, J. R. 1994. Do women with acute myocardial infarction receive the same treatmen as men? *BMJ* **309**:563–6.

Collins, R.; Gray, R.; Godwin, J.; Peto, R. 1987. Avoidance of large biases and large random errors in the assessment of moderate treatment effects: The need for systematic overviews. *Stat. Med.* **6**:245–50.

CONSORT Website. 2001. www.consort-statement.org (accessed 20 January 2002).

Council of Biology Editors' Style Manual Committee. 1994. *Scientific Style and Format. The CBE Manual for Authors, Editors, and Publishers.* 6th edn. New York: Cambridge University Press.

Crichton, M. 1975. Medical obfuscation: Structure and function. *N. Engl. J. Med.* **293**:1257–9.

Crossan, L.; Smith, R. 1996. The BMJ/EASE workshop for editors. *CBE Views* **19**:29–30.

Danel, C.; Moh, R.; Minge, A.; *et al.*, for the Trivacan ANRS 1269 trial group. 2006. CD4-guided structured antiretroviral treatment interruption strategy in HIV-infected adults in West Africa (Trivacan ANRS 1269 trial): a randomised trial. *The Lancet* **367**:1981–9.

Day, R. A. 1995. *Scientific English. A Guide for Scientists and Other Professionals.* 2nd edn. Phoenix: Oryx Press.

Day, R. A.; Gastel, B. 2006. *How to Write and Publish a Scientific Paper.* 6th edn. Westport, CT: Greenwood Press.

De Looze, S. 2002. Rules on virgules. *European Science Editing* **28**:108–10.

Dembiec, D. P.; Snider, R. J.; Zanella. A. J. 2004. The effects of transport stress on tiger physiology and behavior. *Zoo Biology* **23**:335–46.

Dixon, B. Plain words please. 1993. *New Scientist* **137**:39–40.

Dobbel, C. 1938. Dr O. Uplavici (1887–1938). *Parasitology* **30**:239–41.

Ehrenberg, A. S. C. 1977. Rudiments of numeracy. *J. R. Stat. Soc. A.* **140(pt 3)**: 277–97.

Emson, H. E. 1994. Christmas wishes. *BMJ* **309**:1738.

European Carotid Surgery Trialists' Collaborative Group. 1998. Randomised trial of endarterectomy for recently symptomatic carotid stenosis: Final results of the MRC European Carotid Surgery Trial (ECST). *The Lancet* **351**:1379–87.

Federle, M. P.; Cohen, H. A.; Rosenwein, M. F.; Brant-Zawadzki, M. N.; Cann, C. E. 1982. Pelvimetry by digital radiography: A low-dose examination. *Radiology* **143**:733–5.

48 Hours. 1982. Director Walter Hill. Paramount/Lawrence Gordon.

Fremont-Smith, M.; Meigs, J. V.; Graham, R. M.; Gilbert, H.H. 1946. Cancer of endometrium and prolonged estrogen therapy. *JAMA* **131**:805–8.

Gardiner, P. J.; Copas, J. L.; Schneider, C.; Collier, H. O. J. 1980. 2-decarboxy-2-hydroxymethyl prostaglandin E1 (TR4161), a prostaglandin bronchodilator of low tracheobronchial irritancy. *Prostaglandins* **19**: 349–70.

Garfield, E. 1986. The integrated Sci–Mate Software System. Part 2: The Editor slashes the Gordian knot of conflicting reference styles. *Current Contents*, **March 17(11)**:81–8.

Godlee, F. 1996. Definition of "authorship" may be changed. *BMJ* **312**: 1501–2.

Gold, D. R.; Wang, Xiaobin; Wypij, D.; Speizer, F. E.; Ware, J. H.; Dockery, D. W. 1996. Effects of cigarette smoking on lung function in adolescent boys and girls. *N. Engl. J. Med.* **335**:931–7.

Goodman, R. A.; Thacker, S. B.; Siegel, P. Z. 2001. What's in a title? A descriptive study of article titles in peer-reviewed medical journals. *Science Editor* **24**:75–8.

Grüters, A.; Liesenkötter, K. P.; Willgerodt, H. 1995. Persistence of differences in iodine status in newborns after the reunification of Berlin. *N. Engl. J. Med.* **333**:1429.

Gustavii, B. 1975. The distribution within the placenta, myometrium, and decidua of ^{24}Na-labelled hypertonic saline solution following intra-amniotic or extra-amniotic injection. *Br. J. Obstet. Gynaecol.* **82**:734–9.

Halliwell's Film, Video & DVD Guide 2006. 21st edn. Walken, J.; Halliwell, L. New York: HarperCollins.

Hamberg, M. 1972. Inhibition of prostaglandin synthesis in man. *Biochem. Biophys. Res. Commun.* **49**:720–6.

Harris, R. L. 1999. *Information Graphics. A Comprehensive Illustrated Reference.* New York: Management Graphics.

Hartley, J. 1994. *Designing Instructional Text*, p. 31. 3rd edn. London: Kogan Page.

Hearse, D., and the Editorial Team. 1992. Of humour, music, anger, speed, and excuses: Reflections of an editorial team after one year in office [editorial]. *Cardiovasc. Res.* **26**:1161–3.

Heim, S.; Kristoffersson, U.; Mandahl, N.; Mineur, A.; Mitelman, F.; Edvall, H.; Gustavii, B. 1985. Chromosome analysis in 100 cases of first trimester trophoblast sampling. *Clin. Genet.* **27**:451–7.

Helenius, G. 2005. Tissue engineering of blood vessels. (Dissertation.) Lund, Sweden: University of Lund.

Hlava, J. 1887. O úplavici [On dysentery] [Journal of Czech Physicians] Jan. **26(5)**:70–4.

Hodgen, G. D. 1981. Antenatal diagnosis and treatment of fetal skeletal malformations: With emphasis on in utero surgery for neural tube defects and limb bud regeneration. *JAMA* **246**:1079–83.

Holmes, W. 1997. Minimum ethical standards should not vary among countries. *The Lancet* **314**:1479.

Hoyer, L.W.; Lindsten, J.; Blombäck, M.; Hagenfeldt, L.; Cordesius, E.; Strömberg; P.; Gustavii, B. 1979. Prenatal evaluation of fetus at risk for severe von Willebrand's disease. *The Lancet* **2**:191–2.

International Committee of Medical Journal Editors. 2002. *Uniform Requirements for Manuscripts Submitted to Biomedical Journals.* www. icmje.org (accessed 20 January 2002).

International Committee of Medical Journal Editors. 1995. Protection of patients' right to privacy. *BMJ* **311**:1272.

International Committee of Medical Journal Editors. 1985. Guidelines on authorship. *BMJ* **291**:722.

JAMA. 2002. Instructions for authors. Published in the first issue of each January and July and available at www.jama.com

Jha, T. K.; Olliaro, P.; Thakur, C. P. N.; Kanyok, T. P.; Singhania, B. L.; Singh, I. J.; *et al.* 1998. Randomised controlled trial of aminosidine (paromomycin) *v* sodium stibogluconate for treating visceral leishmaniasis in North Bihar, India. *BMJ* **316**:1200–5.

Karman, H.; Potts, M. 1972. Very early abortion using syringe as vacuum source. *The Lancet* **i**:1051–2.

Kartulis, S. 1887. O. Uplavici, Ueber die Dysenterie (review). *Centralblatt für Bacteriologie und Parasitenkunde* **1(18)**:537–9.

Kerkut, G. A. 1983. Choosing a title for a paper. *Comp. Biochem. Physiol.* **74A**:1.

Kesling, R. V. 1958. Crimes in scientific writing. *Turtox News* **36**:274–6.

Kitin, P. B.; Fujii, T.; Abe, H.; Funanda, R. 2004. Anatomy of the vessel network within and between the tree rings of *Fraxinus lanuginosa* (Oleaceae). *Am. J. Botany.* **91**:779–88.

Kurki, T. 1992. Preterm birth. A clinical, biochemical and bacteriological study. (Dissertation.) Helsinki, Finland: University of Helsinki.

Lancet, The. 1993. OCs o-t-c? [editorial] **342**:565–6.

Lancet, The. 1995. English as she is wrote [editorial]. **346**:1045.

Lang, T. A.; Secic M. 1997. *How to Report Statistics in Medicine. Annotated Guidelines for Authors, Editors, and Reviewers.* Philadelphia: ACP.

Laupacis, A.; Naylor, C. D.; Sackett, D. L. 1992. How should the results of clinical trials be presented to clinicians? [editorial]. *ACP Journal Club,* **May/June**:A12–A14.

Lee, A.; Thomas, P.; Cupidore, L.; Serjeant, B.; Serjeant, G. 1995. Improved survival in homozygous sickle cell disease: Lessons from a cohort study. *BMJ* **311**:1600–2.

Lindsay, D. 1989. *A Guide to Scientific Writing. Manual for Students and Research Workers*, p. 36. Melbourne: Longman Cheshire.

Liu, L. 1996. Fate of conference abstracts. *Nature* **383**:20.

How to Write and Illustrate a Scientific Paper

Logan, R. F. A.; Little, J.; Hawtin, P. G.; Hardcastle, J. D. 1993. Effect of aspirin and non-steroidal anti-inflammatory drugs on colorectal adenomas: Case–control study of subjects participating in the Nottingham faecal occult blood screening programme. *BMJ* **307**:285–9.

López-Jaramillo, P.; Delgado. F.; Jácome, P.; Terán, E.; Ruano, C.; Rivera, J. 1997. Calcium supplementation and the risk of preeclampsia in Ecuadorian pregnant teenagers. *Obstet. Gynecol.* **90**:162–7.

Macknin, M. L.; Piedmonte, M.; Calendine, C.; Janosky, J.; Wald, E. 1998. Zinc gluconate lozenges for treating the common cold in children. A randomized controlled trial. *JAMA* **279**:1962–7.

Majewski, J. 1994. Sydsvenska Dagbladet [The South Swedish Daily News], Nov. 6, Sect. A:2 (cols. 2–3).

Marvin, P. H. 1964. Birds on the rise. *Bull. Entomol. Soc. Amer.* **10**:194–6.

Mathews, K. A.; Sukhiani, H. F. 1997. Randomized controlled trial of cyclosporine for treatment of perianal fistulas in dogs. *J. Am. Vet. Med. Assoc.* **211**:1249–53.

McBride, W. G. 1961. Thalidomide and congenital abnormalities. *The Lancet* **2**:1358.

McGarry, G. W.; Gatehouse, S.; Hinnie, J. 1994. Relation between alcohol and nose bleeds. *BMJ* **309**:640.

McNab S. M. 1993. Non-sexist language. *TWIOscoop* **11**(5):148–53.

McWhorter, T.J.; Martínez del Rio, C. 2000. Does gut function limit hummingbird food intake? *Physiological and Biochemical Zoology* **73**(3):313–24.

Medical Research Council. 1948. Streptomycin treatment of pulmonary tuberculosis. *Br. Med. J.* **2**:769–82.

Mehrotra, P. K.; Karkun, J. N.; Kar, A. B. 1973. Estrogenicity of some nonsteroidal compounds. *Contraception* **7**:115–24.

Millar, J. A. 1982. Anonymity of anthropoid apes featured in medical journals. *The Lancet* **ii**:940.

Mills, J. L. 1993. Data torturing. *N. Engl. J. Med.* **329**:1196–9.

Moher, D.; Schulz, K. F.; Altman, D., for the CONSORT Group. 2001. The CONSORT statement: Revised recommendation for improving the quality of reports of parallel-group randomized trials. *JAMA* **285**: 1987–91.

Morell, C. J.; Walters, S. J.; Dixon, S.; Collins, K. A.; Brereton, L. L. M.; Peters, J.; *et al.* 1998. Cost effectiveness of community leg ulcer clinics: Randomised controlled trial. *BMJ* **316**:1487–91.

Mosteller, F. 1992. Writing about numbers, p. 378. In Bailar J. C. and Mosteller, F., editors. *Medical uses of statistics.* 2nd ed. Boston: NEJM Books.

Murray, G. D. 1991. Statistical aspects of research methodology. *Br. J. Surg.* **78**:777–81.

Naylor, A. S. 2005. Differential effects of voluntary running on hippocampal plasticity in the adult rat brain. (Dissertation.) Göteborg, Sweden: The Sahlgrenska Academy at Göteborg University.

Os, J. van; Neeleman, J. 1994. Caring for mentally ill people. *BMJ* **309**: 1218–21.

Pitnick, S.; Spicer, G.S.; Markow, T.A. 1995. How long is a giant sperm? *Nature* **375**:109.

Publications Committee for the Trial of ORG 10172 in Acute Stroke Treatment (TOAST) Investigators. 1998. Low molecular weight heparinoid ORG 10172 (danaparoid) and outcome after acute ischemic stroke. *JAMA* **279**:1265–72.

Quesada, M.; Bollman, K.; Stephenson, A. G. 1995. Leaf damage decreases pollen production and hinders pollen performance in *Cucurbita texana*. *Ecology* **76**:437–43.

Raio, L.; Ghezzi, F.; Di Naro, E.; Gomez, R.; Lüscher, K. P. 1997. Duration of pregnancy after carbon dioxide laser conization of the cervix: Influence of cone height. *Obstet. Gynecol.* **90**:978–82.

Reed, D. M. 1990. The paradox of high risk of stroke in populations with low risk of coronary heart disease. *Am. J. Epidemiol.* **131**:579–88.

Ridley, M. 2003. What makes you who you are. *Newsweek*, **June 2**, 50–7.

Rothwell, P. M. 1995. Can overall results of clinical trials be applied to all patients? *The Lancet* **345**:1616–19.

Samuelsson, S.; Sjövall, A. 1973. Komplikationer och komplikationsprofylax vid gynekologisk laparoskopi. [Complications and prophylaxis in gynaecological laparoscopy.] (In Swedish with English abstract.) *Läkartidningen* **70**:2570–4.

Sarna, S.; Kivioja, A. 1995. Blunt rupture of the diaphragm. *Ann. Chir. Gynecol.* **84**:261–5.

Scientific Style and Format. 1994. *See* Council of Biology Editors' Style Manual Committee.

Spiers, A. S. D. 1984.Transatlantic medical English. *The Lancet* **ii**:1451–3.

Stockdale, T. 2000. Contaminated material caused Creutzfeldt-Jacob disease (CJD) in some undersized children who were treated with growth hormone (GH). *Nutr. Health.* **14**:141–2.

Strunk, W., Jr.; White, E. B. 2000. *The Elements of Style.* 4th edn. Boston: Allyn & Bacon.

Sumner, D. 1992. Lies, damned lies – or statistics? *J. Hypertens* **10**:3–8.

SUN, Xiao-Lin; ZHOU, Jing. 2002. English versions of Chinese authors' names in biomedical journals. Observations and recommendations. *Science Editor* **25**:3–4.

Sundby, J.; Schei, B. 1996. Infertility and subfertility in Norwegian women aged 40–42. Prevalence and risk factors. *Acta Obstet. Gynecol. Scand.* **75**:832–7.

How to Write and Illustrate a Scientific Paper

Synnergren, O. 2005. Time-resolved X-ray diffraction studies of phonons and phase transitions. (Dissertation.) Malmö, Sweden: Universities of Lund.

Thalidomide UK. 2006. www.thalidomideuk.com (accessed 27 July 2006).

The ACS Style Manual. 2002. *A Manual for Authors and Editors.* 2nd edn. Edited by Janet S. Dodd. Washington, DC: American Chemical Society.

Theander, E. 2005. Living and dying with primary Sjögren's syndrome. Studies on aetiology, treatment, lymphoma, survival and predictors. (Dissertation.) Malmö, Sweden: University of Lund.

Tjio, Joe Hin; Levan, A. 1956. The chromosome number of man. *Hereditas* **46**:1–6.

Toogood, J. H. 1980. What do we mean by "usually"? *The Lancet* **i**:1094.

Tønnes-Pedersen, A.; Lidegaard, Ø.; Kreiner, S.; Ottesen, B. 1997. Hormone replacement therapy and risk of non-fatal stroke. *The Lancet* **350**:1277–83.

Truss, L. 2003. *Eats, Shoots & Leaves. The Zero Tolerance Approach to Punctuation.* London: Profile Books.

Tufte, E. R. 1983. *The Visual Display of Quantitative Information.* Cheshire, CT: Graphics Press.

Van Loon, A. J. 1997. Making it easier to trace articles in scientific and technical periodicals: The importance of the first page. *European Science Editing* **23**:9–12.

Vancouver Document. 2002. *Uniform Requirements for Manuscripts Submitted to Biomedical Journals.* www.icmje.org (accessed 20 January 2002).

Vane, J. R. 1971. Inhibition of prostaglandin synthesis as a mechanism of action of aspirinlike drugs. *Nature* **231**:232–5.

Waldron, H. A. 1995. English as she is wrote. *The Lancet* **346**:1567–8.

Wallengren, J. 1998. Brachioradial pruritus. A recurrent solar dermopathy. *J. Am. Acad. Dermatol.* **39**:803–6.

Watson, J. D.; Crick, F. H. C. 1953. Molecular structure of nucleic acids. A structure for deoxyribose nucleic acid. *Nature* **171**:737–8.

Welch, W. J.; Peng, B.; Takeuchi, K.; Abe, K.; Wilcox, C. S. 1997. Salt loading enhances rat renal TxA_2/PGH_2 receptor expression and TGF response to U-46,619. *Am. J. Physiol.* **273** (*Renal Physiol.* **42**): F976–F983.

White, J. V. 1988. *Graphic Design for the Electronic Age. The Manual for Traditional and Desktop Publishing*, p. 201. New York: Watson–Guptill.

Woods, J. 1967. Oral contraceptives and hypertension. *The Lancet* **2**:653–4.

Xu, ZhaoRan; Nicolson, D. H. 1992. Don't abbreviate Chinese names. *Taxon* **41**:499–504.

Index

How to Write and Illustrate a Scientific Paper

Circular graph. *See* Pie chart
Claiming priority, 77
Clemens, Samuel L. – a.k.a. – Twain,
    Mark, 128
Collins, Rory, 117
Column chart, 24
Conclusion, in discussion section, 69
Conference abstract, 62, 121
Confidence interval, 72, 116
Consent for publication of
    photographs, 142
CONSORT, 67
Contraception, 6, 10, 58
Contributors' list, 55, 56
Controlled trial. *See* Clinical trial
Copyright, 145
Correction marks, 139
Council of Biology Editors' Style
    Manual Committee, 107
Credit line, 146
Crichton, Michael, 61
Crick, Francis H. C., 8, 130
Curves in graphs, 20, 45

Data-point symbol, 21
Decimal point, 106
Diacritical mark, 89
Disclosure. *See* Conflict of interest
Dissertation, 91
Dixon, Bernard, 8
Doctoral thesis, 91
Double-blind, 66
Double spacing, 47, 125, 128
Drawings, 38
Dropouts, 69
Duplicate (dual) publication, 143

Ehrenberg, A. S. S., 109
Electronic style, 90
Elements of Style, The, 6
English as a foreign language, 3
Enumeration, 110

Enzymologia, 136
et al., 84
Ethics, 56, 85, 142
Exclusion criteria, 64

Faculty of 1000 Biology, 19
Faculty of 1000 Medicine, 19
Figures. *See* Graphs
First person, usage, 8
Flow charts, 69
Follow-up, 70
Footnotes, 47

Gender, 7
Gibbon, Edward, 3
Golden ratio, 22
Grants, acknowledgement, 80
Graphs
    axis labels, 23, 24, 46
    box-and-whisker plot, 33
    lines, 20
    pie charts, 36
    reduction of, 37
Grouped column charts, 25

Halliwell's Film Guide, 112
Hand-drawn figures, 39
Harvard system, 81, 82, 84
He/she, 7
Headings, running title, 123
Health care, writing on, 130
Helvetica (typeface), 128
Hemingway, Ernest, 17
Heterogeneity, 117
Hyphenation at line ends, 124

Illustration. *See* Graphs
Impact factor, 18
Incidence, 11
Inclusion criteria, 64
Index Medicus, 3, 89, 122, 126
Individual data, 28

How to Write and Illustrate a Scientific Paper